MECHANISMS IN WORLD AND MIND

Perspective Dualism, Systems Theory,
Neuroscience, Reductive Physicalism

BERND LINDEMANN

imprint-academic.com

Copyright © Bernd Lindemann, 2014

The moral rights of the author have been asserted.
No part of this publication may be reproduced in any form
without permission, except for the quotation of brief passages
in criticism and discussion.

Published in the UK by
Imprint Academic, PO Box 200, Exeter EX5 5YX, UK

Distributed in the USA by
Ingram Book Company,
One Ingram Blvd., La Vergne, TN 37086, USA

ISBN 9781845407704

A CIP catalogue record for this book is available from the
British Library and US Library of Congress

Comments and discussion: phblin@uks.eu

Acknowledgement.

It is a pleasure to thank Jürgen Schnakenberg for encouragement and help with physical issues. His benevolent advice led to a considerable shortening of the manuscript.

From discussions with Rüdiger Brennecke I learned essentials about causation.

The author

is Professor for Physiology (retired) at the Medical Faculty, Universität des Saarlandes, Germany.

Contents

1	Perspective	1
2	Mechanisms	9
3	Systems	27
4	Reductive modelling	41
5	Neuronal systems	63
6	Mind	79
7	Consciousness	101
8	Summary	113
9	Facit	123
10	Bibliography	128
11	Glossary	137
12	Index	148

1. Perspective

This text on philosophical aspects of neuroscience is centred on perspective dualism, distinguishing the mental or first-person view from the neuronal world, which is invisible to the first-person, the Self. The topic *reduction of apparent mental processes to real neuronal mechanisms* unfolds in the discussion of mechanisms in world and mind. Models of neuronal mechanisms of differing complexity are described in a general way, classified and assigned to levels of systems theory. Various strategies of reduction are delineated and their feasibility is tested using explananda such as life, mind or consciousness.

The aim is to explore if and how the mental may be understood in terms of neuroscience, in terms of biophysical mechanisms. According to a common intuition, such understanding is not possible because humans have design, agency – they have feelings, emotions, consciousness - and intention, concepts, knowledge, reason, believes, values, dignity - they are in many ways 'more' than what is explainable by physics. Nevertheless, such understanding in neuroscience terms will be found feasible for a variety of one-level reductions, including reductions of agency and dignity. Multi-level reductions are combinations of one-level-reductions. Their explanations, unfortunately, are less comprehensible, for comprehension tends to fail as intermediate explanations are skipped.

1a. The mind-neuron problem

When viewing living beings from a distance, in the third-person perspective, we find them to be similar to physical objects in a general way. For they can be located in space and time (are not abstract) and events within them, like those outside of them, follow gap-less chains of physical interactions. Thus the living objects, or their bodily aspects, are part of the physical world. As such they can be analysed objectively by physics and its branches biophysics, biochemistry, genetics, molecular biology, neurobiology, biology. This third-person view is the *physical perspective*.

But there is something else. Living beings, unlike physical objects generally, have design, a construction plan generated and changed by evolution. Further, they have needs and initiative by design. For instance, when their composition deviates too much from optimal values, living beings take measures of self-preservation, counteracting the unfavourable trend.[1] Then they will fold or unfold their leaves or tentacles, expand or shrink, search for food, leave their environment or manipulate it, in short, make use of mechanisms of control to exercise autonomy, agency.[2] This property of agency draws a line between them and their environment, it establishes a subjective *agent-world polarity*.

Further, those living beings which live with peers *report* about themselves by their behaviour and those gifted with speech *report* about themselves explicitly, using a system of symbols.[3] Their story reveals an inner dimension which seems fundamentally different from the physical world. Using the first-person or *mental perspective*, the reports are about experiences, thoughts, desires, about feelings of a conscious Self-agent. The reports are about mind-phenomena which are not locatable in space (and, arguably, time). Where in space is a thought, a belief? Being not

---

1 This behaviour may be shared by automata designed by humans.
2 An agent is an at least partially autonomous unit which is or appears to be in control: it can decide and act upon its decisions. This definition applies, for instance, to every living cell.
3 The report is an objective fact while its content is only subjectively accessible.

locatable, mind-phenomena are abstracta. As such they cannot interact with physical things, even though they are 'about' them [21]. Yet such thought → world interaction is reported by the Self-agent, a logical contradiction.[4]

Mental and physical perspective, first- and third-person perspective, these are the two sides of 'perspective dualism'[5]. The perspectives may be equally relevant but they are contradictory and irritating: Why two points of view? Below I shall try to answer this vexing question.

It comes almost as a relief that there is a clear relation between the mental and the material world. The mental has a support system, the body. In detail, experience shows the mind's objective existence to depend on many organs but in particular on the action of the body's neurons.[6] When influencing the neurons physically, the mind is strongly affected. For instance, when, due to a physical effect on neurons, consciousness is lost, the mental stops to be noticeable. When the neurons cease to interact entirely and life ends, any report from this mind ceases too. Whether the mind continues to exist is a matter of belief. But the familiar report, which was an objective fact, ends with the life.

Thus there is a dependence of mind on neurons. Whether there is in addition a dependence of neurons on mind is the other

---

4   'An object is abstract if and only if it is non-spatial and causally inefficacious.' http://plato.stanford.edu/entries/abstract-objects/ , see also the *fallacy of mistaken concreteness* [139:51].

5   I use the term *'perspective dualism'* to designate 'first-versus third-person perspective'. The concept is also known as epistemic dualism [57, 58]. Contrary to J. Habermas I do not imply that the mental perspective necessarily cannot be reduced to the physical. For roots of perspective dualism and the mind-body problem, see René Descartes, quoted in [112:142], Franz Brentano [21:124], Ludwig Bertalanffy [16:95ff] and Thomas Nagel's 'dual aspect theory' [101:28].

6   Arguably the mental 'supervenes' over the neuronal, meaning that it exists 'in virtue of', is necessitated by the neuronal system. Supervenience [35] is an asymmetric relation: the supervenient cannot change without a change in the subvenient. According to John Heil, the relation may be identitiv, constitutive or causal [60:67]).

question of the mind-neuron problem. It will be taken up.

A mere dependence of mind on neurons would not prove that neuronal activity *alone* 'gives rise to' and explains the mind. Yet it opens this possibility. Already the 17th century 'philosophical materialism' [7] claimed dependence, that "all that exists is matter in motion and mental states are ontologically dependent on states of bodies" [64]. Around 1820, *"organ physics"* was the inspiration of a group of influential physiologists. Opposing vitalism and nativism, Helmholtz, together with du Bois Reymond and others, aimed to reduce the phenomena of the living body to mechanisms based on chemical and physical laws, cast in mathematical form [e.g. 23]. This third-person approach, still largely excluding the 'mind-body' problem, was a resounding success.

Physicalism,[8] which pointedly deals with the mind-body or mind-neuron problem, is still a controversial branch of philosophy. A defender of *reductive physicalism* (RP for short) expects that mental activity can or will be explained by biophysical neuronal processes. Indeed, according to Jaegwon Kim almost all mental states, excepting only the qualia, are reducible to neuronal (in the end, physical) processes [74]. In contradistinction, a defender of *post-reductionism* maintains that such reduction is not possible as the reduction base is incomplete and the mental more than a physical system.

The promise of reductive physicalism is a unity of science. One world, one science, including the expectation that the mental first-person perspective will be explained by neuronal phenomena. RP claims that even psychology or sociology or ethics will, though indirectly, have a physical basis. I add that such multi-level reduction to physical base is, where possible, of little practical use. For the weakness of RP becomes apparent when many system levels are included in the reduction. Then the explanatory appeal decreases for reasons to be explained. As reduction progresses, comprehension fails.

---

7 also known as 'ontological physicalism' or 'physical monism'.
8 *Physicalism:* All is physical or supervenes on the physical. All phenomena can be explained physically.

1b. Three concepts

World, mind and mechanisms are key concepts of this treatise. The *physical world*, of course, is first of all our environment, a system of objects of matter,[9] composed largely of atoms. These, in turn, are composed of subatomic particles, which are composed of elementary particles and/or waves. Every change in this physical world is due to interaction of matter and its constituents, their energy and fields (Section 2b). The world includes ourselves as physical entities, our neuronal mechanisms are mechanisms of the world.

The human *mind* is a bundle of *experienced mental processes*, tentatively taken to be a result of neuronal (physical) mechanisms in our brain. These mechanisms arguably generate the *first-person perspective* as the experience of an interior view. The view shows our conscious Self positioned in the world in past, presence and future. The first-person or Self views itself to exist apart from the world in an agent-world polarity. Further, it is not aware of its own neuronal system. This because we perceive only what our senses tell us, and we have no sensory organ to notice our own neuronal activity.

In a way the Self is a *separation by perspective*, an agent experiencing independence from the world and from its organism's machinery. The separation is of great conceptual consequence. Of course it would be illusionary to conclude that Self and mind actually have this self-perceived independent existence. Indeed, the objective third-person perspective shows Self and mind to depend on the action of neurons in the physical world.

A *mechanism* is a physical device or system (made of physical components) optimized to alter its environment in a characteristic and quantitatively more or less predictable way.[10] Mechanisms are designed by man or have evolved in nature. They are

---

9 'Matter': what has mass and other classical physical properties (Extensive: energy, mass, charge, volume. Intensive: speed, temperature, pressure, density and others). Photons, being without mass, are not part of matter (e.g. http://en.wikipedia.org/wiki/Matter). They, of course, are part of the world, too.

10 For further definitions, see [10:13ff].

modelled with physical cause-effect chains (causal chains or causal loops), defining the sequence of component interactions giving rise to state transitions. The design appears to be *optimized* to yield a distinct system behaviour (SB) ranging from random to almost fully predictable. The study of neuronal mechanisms, ordered by systems theory, may open the way for an understanding of the mental in neuronal terms.

1c. Preview

(1) **Neuronal mechanisms**, though built of molecules with stochastic behaviour, are often modelled deterministically. One thread of this text concerns the *idealizations* which lead to such deterministic models. The (usually macroscopic) models describe only the mean values of fluctuating ensemble sums. However, full deterministic reliability is not possible for neuronal mechanisms, as probabilistic fluctuations, which even increase with ensemble size, are ubiquitous.

In models of neuronal mechanisms a *causal chain* and a *causal loop* is a characteristic sequence of physical interactions. These are more or less optimized, raising the efficiency of the system behaviour SB. Even probabilistic models are causal and there is no evidence for 'causality-gaps' in the so-called 'causal closure of the physical world'. Thoughts, if understood as immaterial, cannot influence material neurons. And there is no need for this influence, because thoughts necessarily have a neuronal basis, and this interacts causally.

(2) **Universal system levels.** Another major theme is the nature of system levels and their linkage. A novel concept is suggested: that each level is *universal*, housing all basal objects and events (which are physical by axiom). These are grouped differently and symbols are assigned to the groups, populating higher-order levels. Events involving basal items occur synchronic on all levels. Linkage of levels is given by identity of the basal items. There is no need for an additional vertical linkage of levels, be it causal or constitutive. The levels importantly differ in the elements' grouping, representation with symbols, thematization of features and in level-specific idiom.

1. Perspective

(3) **Reductive physicalism.** Reduction, the attempt to explain by fundamental laws, is a recursive process for which several examples are given. The reductions of life, of mind and of consciousness to physical processes without remainder are such attempts. Who stops his reductions abstains from any explanation.

Reductive physicalism brings the expectation that the mental will be understood in neuronal and thus in physical terms. This seemingly daunting task may be aided by systems theory. It is common to place a mental level above the neuronal. But the nature of those levels is important. With *universal* levels basal events are physical per axiom, the physical basis of all complex phenomena is implied. This arguably includes the phenomena of the mind.

Reductive physicalism cannot be proven, it is a hypothesis. As yet, the hypothesis was not falsified, though often rejected. It rests on the causal closure hypothesis of physics. Mental functions like thoughts, if immaterial, must be causally powerless. For they are *'about'*, are language, abstracta, they cannot encounter for physical interaction. The causal relevance, which the mental nevertheless appears to have, may be due to its neuronal roots which, as part of the physical world, have causal power.

Generally mental (immaterial) phenomena have an objective physical reduction base over which they supervene. That, in short, is the RP hypothesis. However such reduction, if it includes multiple levels, explains little in a 1st-person perspective, it tends to over-tax our comprehension. That is the devil's hoof. The mental, which is 'more' than physics, may be reducible to physical base by objective scientific proof. Yet subjectively it will not be convincingly explained by physics because such explanation exceeds our cognitive abilities. Proof is objective, comprehension subjective.

> 'More' than physics: yes.
>
> Objective reduction to physical base: yes.
>
> Subjective comprehension: limited.

2. Mechanisms

Mechanisms are defined and examples are given for their probabilistic and deterministic models. When based on the behaviour of molecules, probabilistic models are more realistic and deterministic models (on the macroscopic level) are their idealisations. Mechanisms operating continuously have cyclic state-transition diagrams and show the over-sum effect. All models of mechanisms are causal. Those of neuroscience are based on molecules, often in small numbers. They show ubiquitous fluctuations and cannot perform ideally deterministic.

A mechanism is a physical device made of interacting components, optimized to alter its environment in a characteristic and quantitatively more or less predictable way by its system behaviour SB. Mechanisms were designed by man or have evolved in nature. In designed mechanisms SB is the construction goal. A machine is constituted of mechanisms. The optimization with respect to SB improves the reliability of mechanisms and machines.

An *ideal* mechanism is a deterministic system of components and their interactions. Its behaviour is fully predictable by appropriate differential equations. In a wider scope the behaviour of mechanisms ranges from probabilistic to deterministic. Typically, molecular models involving few units behave probabilistic, models of large ensembles of molecules behave quasi-deterministic and macroscopic models may behave deterministic.

2a. Examples and basic features

The interaction of components of ideal mechanisms may be modelled kinetically, e.g. with a set of ordinary differential equations with continuous variables [e.g. 17, 91]. The differential equations are deterministic.

The equations of finite state models define system states and specify the rate of change of their occupation, which may be probabilistic. The states are exclusive and, together with their transitions, form a structure given by a *state-transition diagram (STD)*. This structure describes the organisation which makes a concerted action of components possible.

Examples for macroscopic single-shot mechanisms are the ignition and burning of a match or the firing of a rocket. The energy set free in such events is limited by an inherent chemical reservoir and there is no recovery of the initial state, a steady state is impossible. In contrast, macroscopic mechanisms and machines which are capable of repetitive (steady-state) performance have cyclic STDs and must tap environmental energy gradients (Figure 2.1A).

Similarly, a cyclic *molecular mechanism* allows repetitive (steady-state) performance, modelled with a cyclic STD [e.g. 63:5]. The cycle assures recovery of the initial state, tapping an environmental energy source (Figure 2.1C). Examples are many: metabolic cycles in cells, molecular ion pumps and ion channels in cell membranes, drug receptors, etc.[11] In molecular models the continuous variables are occupational probabilities: the equations specify these system state probabilities and their rate of change (finite state Markov-chain models). Typical of

---

11 Usually unconsidered in such models remains the fact that repeated action leads to ageing of components and thus to a slow change of the mechanism itself. Then real recovery is an idealisation. In biological cells there is continuous replacement of ageing components by newly synthesized ones.

2. Mechanisms

molecular mechanisms is the probabilistic system behaviour.

Constituted of molecular devices we find a variety of *neuronal mechanisms* of higher organisation like synapses, neurons, reflex circuits, further neuronal feedback circuits and regulators, analysers of sensory data, memory devices, conscious-executive cortical mechanisms etc. Their output is largely driven by the input. Details are provided by a wealth of deterministic and stochastic models in neuroscience [e.g. 78, 80].

The *brain* is an organ devoted to the control of body function, analysis of sensory input, coordination of muscle activity, generation of emotions and wishes, conscious thinking, conscious experiencing of past (memory), present and predicted future and many other tasks. These complex functions are all due to interaction of neurons, which are built of stochastic elements, molecules. The complex functions combine in various ways, generating internal states and behaviour performed in the environment. The brain's neuronal activity may be described as ideally deterministic only if its result is fully predictable. While, strictly speaking, this will never be the case,[12] the measured performance may still be impressively predictable in selected cases.

Every mechanism has its specifications and, therefore, limitations. Regarding the human brain as an assembly of neuronal mechanisms, several limitations are apparent. The restricted capacity of working memory [96] is certainly one of them (see PLC, Sections 4j, 5c).

In conclusion, there is a hierarchy of organisation concerning mechanisms in neuroscience, with models spanning from molecules to neuronal networks and brain modules. On the molecular level the model performance is probabilistic or, in large ensembles, quasi-deterministic. At higher levels of organisation it may be nearly but not fully deterministic. For, based on molecular components, a residual lack of certainty cannot be excluded.

---

12 if based on a level-2 formalism (mesoscopic physical level, see Section 2b).

Term	Meaning
Mechanism	Optimized physical system with typical, ideally with predictable behaviour
Component	Of a mechanism. Distinct physical entity capable of interaction and energy transfer
State	Of component or system. One of several mutually exclusive combinations of component properties as defined in the STD
STD	State-transition-diagram. Defines all states and transitions
Transition	Event of changing state. Due to time-consuming interaction of components
OCA	Organized component activity. Transitions as defined in the STD
Cycle-rate	Cycle-turnover. Rate of repetitive cycle-completion
SB, EB	System behaviour. Cycle-rate or its direct consequence, the environmental behaviour
MaW	'Mechanism-as-a-whole' [30]
D- or P-model	Deterministic or probabilistic model
1B2S	1-barrier 2-site ion channel

Table 2.1. Key terms: Abbreviations and meaning.

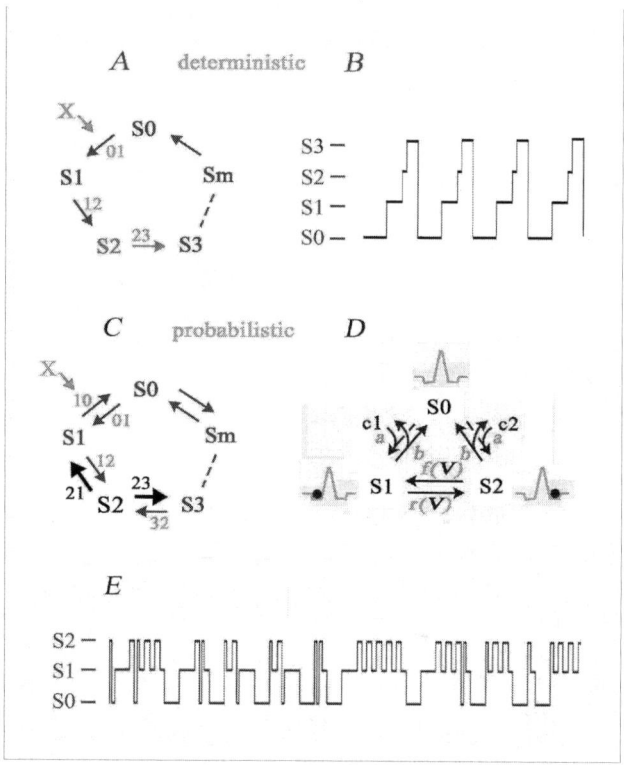

Figure 2.1. *A:* State-transition diagram of a *deterministic model* with environmental energy supply X. Only one path leads away from anyone state. *B:* Time course of transitions of a 4-state deterministic model. *C:* State-transition diagram of a *probabilistic model* showing a single multi-state cycle. S2 can be terminated in two ways (bold arrows). Numbers are indices of transition-rate constants k. *D:* STD of a 1B2S (1-barrier 2-site) ion channel. *E:* Time course of transitions of a 3-state model of one channel molecule, obtained by Monte Carlo simulation. Such random behaviour of single units occurs, of course, even in the steady state of large ensembles.

2b. Levels of description in physics:

At least three domains of description are commonly distinguished. Here I follow a scheme of levels 1-3 communicated to me by Prof. Jürgen Schnakenberg, Aachen. I deviate from it merely by splitting level-1 into a microscopic and a basic level.

3. Macroscopic level: By evaluating the mean values of probabilistic descriptions of level-2 and neglecting fluctuations, **deterministic** solutions of 'phenomenological' differential equations for the *means* of the variables are obtained.

2. Mesoscopic level: By focussing on a few representative variables of level-1, solutions may be obtained as **probabilistic** descriptions. The stochastic, fluctuating behaviour arises from a neglect of parts of the microscopic information of level-1.[13]

1. Microscopic level, describing complex objects still constituted of **matter,** having mass and other classical physical properties (Extensive: energy, mass, charge, volume. Intensive: speed, temperature, pressure, density and others). Description of phenomena in many cases by **deterministic** differential equations. These, generally, are too complex to be solved. Yet it is likely or known that solutions exist.

0. Basic or elementary (quantum-physical) level, describing subatomic particles (electrons, positrons, neutrons), of which positrons and neutrons are constituted of various quarks. Electrons and quarks are **elementary particles** which have no constituents. Level-0 further describes photons and quantum effects, and the complements of elementary particles and photons: **wave packets** (probability amplitude waves). - I gather that level-0 is highly probabilistic [146]. The familiar concepts of matter, time and space, of causation and interaction do not apply. According to Anton Zeilinger the basic concept

---

13 Thus two coinciding events (each causally explained) may seem independent on the mesoscopic level, their coincidence apparently due to chance. Yet this independence does not hold on the microscopic level, chance on level-2 is due to a neglect of level-1 information.

of level-0 reality is instantaneous information [145]. - Subjective comprehension fails entirely.

Note that many microscopic events are taken to be deterministic, on level-1 objective chance is resolved by taking full account of the complexity. Indeed, determinism is for many systems on level-1 quite likely or even certain, when it is known that solutions of the model equations exist. In contrast, the quantum-physical world is *not* deterministic, we cannot predict single events. Thus objective chance exists on level-0 [e.g. 135:32, 146]. Only the behaviour of large ensembles can be predicted, a parallel to level-3..[14]

The steps from level 1 → 2 and 2 → 3 are achieved by a neglect of level-1 and level-2 information. We may suspect that such neglect is, at least in part, necessitated by epistemic constrains: our difficulty to mentally handle large systems [e.g. 96]. Thus truly realistic descriptions would be those of level-1, while descriptions of level-2 and -3 trade strict realism for the ability to solve the equations and comprehend the results. Against such difficulties scientists strive for objective descriptions with models approximating the reality of level-1 and level-0.

The text below is concerned with level-2 and -3. Probabilistic models belong to the 2nd physical level and 'phenomenological' deterministic models, derived from them, to the 3rd level. However, as will be shown, historically the sequence of discoveries was often 3 → 2 rather than 2 → 3.

2c. Model types in detail:

Consider a general flow(force) relationship, where J is flow and X is force:

$$J = g^* X.$$

When viewed with conventional systems theory, g^* is at system

---

14 Where interaction with the environment results in decoherence, usually in large systems, quantum-effects based on coherence are not likely to be observed. See http://en.wikipedia.org/wiki/Quantum_decoherence.

level n a black box of measured but unexplained properties, it is the explanandum. When viewed at level n-1, g* is explained by (reduced to) a distinct conductance model specifying the mechanism. Depending on the measured properties, this may be

1. a macroscopic, deterministic model specifying homogeneous material properties or

2. a macroscopic model describing components interacting deterministically. Alternatively, it may be

3. a mesoscopic model specifying an assembly of molecules behaving probabilistically. Neuronal mechanisms are often small, constituted of few molecules. They show pronounced fluctuations of single-unit behaviour and of multi-unit system behaviour. Then predictions of neuronal SB become less reliable.

Thus, with respect to their reproducibility two major types of models of mechanisms, probabilistic and deterministic ones, will be discussed. They belong to the 2nd and 3rd physical level.

D-model: Ideal *deterministic* models of mechanisms are those where repeated "trial runs" yield identical solutions. The solution of one run predicts the solution of the next run reliably. This is the case for finite-state models with macroscopic components, which show the same sequence of states when repeated runs start with identical states. There are no random choices because each state is terminated by a transition to just one possible succeeding state, as specified in the STD. This diagram shows at least one unidirectional cycle to allow steady state performance (Figure 2.1A), resulting in a fixed sequence of interactions (Figure 2.1B). The lifetime of a state S_i is related to the inverse transition rate constant k_{ij}.

The time requirements are due to macroscopic phenomena such as filling of a reservoir, movement towards a contact point, rise of voltage to a threshold value or pulse propagation to a distant receiver. An environmental energy source is required.

Due to the absence of random choices the SB (the cycle rate) is identical in repeated runs, it does not fluctuate. Such perfect reliability is not possible with molecular components.

2. Mechanisms

P-model: *Probabilistic* finite-state models typically describe the behaviour of molecules. Fluctuating thermal energy exceeds a barrier with a certain probability, thus terminating state S_i by transition to S_j.[15] This process is partially balanced by transitions from S_j to S_i. Then states can be terminated in two ways. The probabilities based on two rate constants describe what happens, the lifetime histogram of S_i shows two peaks related to $1/k_{ij}$ and $1/k_{ih}$. The resulting sequence of transitions is random, but within constrains, since each choice is between but two subsequent states (Markov chain), as specified in the STD (see S_2 in Figure 2.1C).

Due to the random choices, different sequences of states are generated when repeated trial runs start with identical states. Thus the net cycle rate (and SB) is quantitatively different (fluctuates) in repeated runs.[16]

However, the SB of mechanisms remains qualitatively predictable, even though not quantitatively the same. For instance, while a single potassium channel may pass potassium ions in or out of a cell with a rate fluctuating unpredictably, it will nevertheless always pass potassium ions in or out of the cell, and this with a mean rate which can be calculated. The SB of the probabilistic model is qualitatively fixed. If SB is not realized qualitatively, the bounds of the model were transgressed, the model abandoned for another one.

In populations of N independent, identical molecules the net cycle rates add to a fluctuating ensemble value J_N which increases with N. On adding the individual rates, the sum of fluctuations also increases with N (though sub-proportionally). The fluctuation amplitude, the population sum J_N and its temporal or ensemble mean value may be obtained numerically, using Monte-Carlo simulation.

Quasi-D-model: While molecular mechanisms are probabilist-

---

[15] In Eyring's transition state theory the 'free enthalpy of activation' is randomly exceeded [44]. Molecular collision frequency is very large, in the order of 6×10^{12} / s.

[16] For details and equations see [63:11-15].

ic, their ensemble-behaviour becomes deterministic, or nearly so, if the population of parallel units is sufficiently large. This is the case for many populations of identical protein molecules. The large number of molecules and their large frequency of thermal vibration allows the formulation of quasi-deterministic process equations based on ensemble means and an additive noise term. By neglecting the noise, the equations become deterministic [e.g. 63, 80, 130, 134].

The quasi-D-model is derived from the P-model by summing the cycle rates across a large ensemble of molecules and thematizing the temporal ensemble mean of J_N while omitting the fluctuations. The model predicts near-deterministic at large N. Yet its predictions are not fully reliable because with increasing N the added fluctuations do not vanish but increase, too. Thus the equations describing the explanandum may become deterministic while the explanandum does not.

2d. Macroscopic deterministic models replaced by molecular probabilistic models. Examples with conductances:

Conduction models describe material properties which, in the macroscopic deterministic case, may be represented simply by a *material constant* multiplied with a geometry factor. Their product g* could be the conductance of a flow(force) relationship, for instance of a current(voltage) equation. The constancy of this conductance at a given force or voltage makes the equation deterministic. Here are examples using a minimum of mathematical formalism:

1. A general form of current(voltage) relationships is

$$I_c = g_c (V-E_c),$$

in which I_c is the current of charge carrier c, E_c the equilibrium potential[17] of the charge carrier and V the voltage across the conductor. The conductance g_c or the current is the explanandum.

---

17 Nernst potential

2. Mechanisms

In metallic conductors the concentration of charge carriers (electrons) is very large and their gradient small, decreasing E_c to negligible values. Ohm's law $I_c = g_c V$ results. Historically this relationship was implicitly explained by a deterministic macroscopic model in which g_c was a proportionality *constant* [18] [106:36]. However, later measurements of metallic conductors have shown g_c to fluctuate at constant V [e.g. 37]. This required a probabilistic model as the reduction base.

2. Another example, relevant to neuroscience, is the equation for the resting voltage V_{oo} of a cell membrane. We start with the current(voltage) relations of only two permeant ionic species in the stationary state:

$$I_{Na, oo} = g_{Na} (V_{oo}-E_{Na}), \quad I_{K, oo} = g_K (V_{oo}-E_K)$$

in which V_{oo} is the membrane voltage at rest, E the equilibrium potential of the ionic species Na^+ and K^+ and I_{oo} the stationary current of the ionic species. g stands for the conduction mechanism. As the two currents are equal and opposite at rest, we obtain

$$V_{oo} = (g_{Na} E_{Na} + g_K E_K) / (g_{Na} + g_K).$$

The underpinning conductance model was originally deterministic: resting g_{Na} and g_K were taken to be constants and so was V_{oo}. Later the advent of membrane noise analysis revealed that V_{oo} was fluctuating in the sub-millivolt range [e.g. 133]. A probabilistic model was needed to explain this, specifying the switching current noise of an ensemble of ion-conducting molecular channels in addition to the current fluctuations of open channels [e.g. 119].

3. A third example from neuroscience is provided by the Hodgkin-Huxley equations. They describe the evolution of ionic currents flowing through voltage-dependent conductances during an action potential:

$$I_{Na} = g_{Na} (V-E_{Na}), \quad I_K = g_K (V-E_K), \quad \ldots$$

in which the coefficients g symbolize conduction mechanisms

---

[18] At least for small ranges of V. The (temperature dependent) *material constant* is the electrical resistivity ρ [ohm meter] with g_c = area / length / ρ.

which change with V and affect each other by the changing V. I, g and V vary with time, but have single values at any time. In the model the values are identical in repeated trial runs, they do not fluctuate. For details, see [65].

The Hodgkin-Huxley model, operating with macroscopic conductances,[19] may be understood as ideally deterministic. However, subsequent measurements at neurons revealed ubiquitous fluctuations of the ionic currents in the nano- and pico-ampere range. They were interpreted with probabilistic models which specified a population of voltage-gated molecular channels replacing the macroscopic conductances [e.g. 119].

The examples emphasize that the deterministic models first invoked, specifying macroscopic conduction mechanisms, did not account for fluctuations. To explain the fluctuations, a change to mesoscopic, probabilistic models was needed. The probabilistic molecular models are more general, applying to small as well as large ensembles, and provide a better reduction base for the explanandum.[20]

As mentioned, in the neuronal system, where functional units are often small, constituted of a limited number of molecules, fluctuations are ubiquitous. The fluctuations are described to have their advantages [e.g. 45]. For instance, as C. Koch pointed out, the reliability of neuronal mechanisms appears to be tuned, in part by varying ensemble size. Low reliability is not an unavoidable fact, but a design feature which increases the dynamic range [78:327].

---

19 even though the possibility of conducting carrier or channel mechanisms was mentioned in [65].
20 In this sense the title "Mythos Determinismus" of Falkenburg's book [46], when applied to the macroscopic determinism of level-3, seems justified. When considering cases where full account of the complexity on level-1 results in determinism on level-2, the title does not apply. On level-0 it is justified again.

2. Mechanisms

> IDM: Idealisation in deterministic models.
>
> In the neuronal system, quasi-deterministic and deterministic models are idealisations focussing on temporal mean behaviour of an ensemble of N units, neglecting fluctuations. Yet fluctuations, though not thematized in D-models, do not vanish but even increase with ensemble size. Fluctuations remain irrespective of our models, the prediction of means can be misleading, full reliability of model predictions is not possible.

2e. Causal models

We know from experience that, where investigated:

> Every change in a physical entity was due to *interaction* with another physical entity.[21]

This fundamental regularity, which is in accord with Newton's third law, may be extended to an expectation or hypothesis. It was, as yet, never falsified.[22] Further, taking any physical change above level-0 as *caused* by an interaction, one arrives at a sequence of causes and effects without interference from outside physics. Thus we model the physical world as 'causally closed', since:

> Every phenomenon which had a cause, had a physical cause.

As Aristotle already observed, some events appear to lack a cause. They are commonly attributed to absolute chance [e.g. 135, 146]. But is it a lack of cause? If I throw a die repeatedly,

---

21 On level-1 and above, but excluding level-0, where *interaction* in space and time does not apply.

22 Atomic decay is sometimes quoted (idiom of level-1) to show change without interaction. However, the decay is due to interaction of certain subatomic entities (nucleus and photon-field), while interaction time is not predictable, appears to be due to chance.

number 6 will come up, **caused** by my repeated throwing. Only the time when it comes up (or the number of unsuccessful trials) is not predictable. Here chance results from a cause lacking predictability with respect to time of effect.

In the light of the above scheme of physical levels of description (Section 2b), much of level-1 is taken to be deterministic, not compatible with 'events without cause'. For in this light chance on level-2 arises from omission of level-1-information. Indeed, when the full complexity of the die-throwing experiment is accounted for, the result is predictable.

On lower levels quantum-mechanical chance plays a decisive role [146]. Physicists cannot causally predict single quantum events, but they can predict the behaviour of large ensembles. Thus "large" ensembles are deterministic-causal. This argues for causality in the supra-atomar world.

Note that by defining chance as the absence of cause [135:34], or the apparent absence of cause, causation and chance are introduced as a pair, we cannot understand one without the other. Indeed, in the supra-atomar world chance events are not all there is, for causal events, causal chains and networks abound. In this world the causality concept seems indispensable.

David Hume called attention to the mental habit of associating two events as cause and effect. Such association happens without sufficient reasoning, as in *post hoc, propter hoc*. Relating cause to effect may be a principle of the human mind rather than of physics and reality [70], see also [e.g. 113:2, 117]. Other authors uphold causation as a valid principle of physics[23] and philosophy [e.g. 52, 55, 71, 142, 143: 67].

Why is Hume's mental principle of associative causation useful? It may have evolved because it replaces the mental representation of a complex network of interactive relations with a simple scheme of two, releasing working memory from an overload of detail (PLC), while still approximating reality.

For sure, in every-day situations we all take cause-effect relations for granted. Often they are applied to neuronal mechan-

---

23 http://en.wikipedia.org/wiki/Causality_(physics)

isms, too. For instance, W. Bechtel and C.F. Craver emphasize that within a neuronal mechanism the components interact causally [10, 30, 31]. Following J. Woodward, C.F. Craver and others, we shall adopt causation as a useful principle at least for the modelling of supra-atomar physical mechanisms.

In the world events may simply happen, or may happen in a large network of deterministic relations. Yet ours is the notion that they happen *'because'*, that there is one direct cause. Such every-day causal models may be successful since they capture a chunk of reality. In any case, in the idiom of level-1:

> Physical causation requires interaction. A *causal chain* is simply a sequence of interactions. In a mechanism the sequence may be *optimized,* raising the efficacy of SB.

c-criteria: Hence we model interactions within mechanisms with cause and effect. We now ask whether a particular event may be modelled as the direct cause of another. To decide this requires experimental intervention. Given the correlated physical events e_1 and e_2, with e_1 preceding e_2, the following scheme of interventions will pragmatically *test for causation*:

With the path $e_1 \rightarrow e_2$ isolated from its environment or while the environment is monitored for changes, a deliberate step-change in e_1 is effected. This may result in an associated but (due to interaction) delayed change in e_2. In a given time interval the change in e_2 may occur with a certain probability only (probabilistic molecular models), nearly always (quasi-deterministic molecular models) or always (macroscopic models).

If the change in e_2 is not explained by associated events in the environment, and if energy or another conserved quantity is exchanged[24] in the transition $e_1 \rightarrow e_2$, then we form the working hypothesis that e_1 is the direct cause of e_2 or is connected to e_2 by intermediate events forming a chain of direct causes. The hypothesis is a causal model.

Notably, while a deliberate change in e_2 may result in an associ-

---

24 The transfer of energy (or of another preserved quantity like mass, momentum, charge) occurs between the components which interact causally. It is not the event (transition) which provides or accepts energy.

ated change in e_1, this must be mediated by a pathway different from the one just discussed. For if testing shows that e_1 is the direct cause of e_2, then e_2 is not the direct cause of e_1. Inversion of pathway $e_1 \rightarrow e_2$ is not allowed (manipulatory asymmetry).

Applying this reasoning to mechanisms, the activity of one component corresponds to e_1 and the activity of another component and ultimately of SB to e_2. Additional pathways through the environment may exist (feedback) but are carefully excluded from the mechanism itself by design or evolution. This makes deterministic models of mechanisms pass the causation test.

Further, since the transition $e_1 \rightarrow e_2$ may occur with a probability < 1, even probabilistic models pass the test. If the sequence of interactions is one of random choices between two possibilities (e.g. Markov chain), then such probabilistic choice can in retrospect be understood as the result of distinct physical causes (like thermal vibrations). In retrospect the direct cause for a stochastic event was that a random variable finally exceeded a threshold value.[25] Also, the randomness may be understood deterministically by taking full account of level-1 information.

In conclusion, D- and P-models fulfil the c-criteria. Thus, not surprisingly, **all models of mechanisms are causal models**.

2f. Causation in cyclic STDs

The characteristic feature of molecular mechanisms performing continuously is the cyclic STD. Closing of the kinetic cycle causes a brief transient, then the rates of all transitions become identical. To calculate the steady state turnover (net cycle-rate), anyone pair of transition rates may be picked and their difference evaluated with established methods. The cycle-rate equation of both deterministic and probabilistic models shows a dependence on all rate-constants of the cycle. Thus the joint, con-

---

25 Here chance is taken not as the general absence of cause [e.g. 135:34] but as the absence of a cause with temporal predictability. Level-2 chance is an artefact arising from omission of level-1 information.

certed activity of all components and transitions is required, single transitions are necessary but not sufficient.

Over-sum: As mentioned, the cycle-rate is dependent on *all* transitions (in the combination given by the right hand side expression of the cycle-rate equation), and *only* on these. Each transition contributes as necessary but not sufficient for the result, a special case related to a contributory cause but not identical to it. In principle, the contributory cause of a transition may be neither necessary nor sufficient. In mechanisms the contribution of a transition is necessary but not sufficient.

Since the cycle-rate depends on all transitions, on the whole mechanism, it will indicate the mechanism's system behaviour SB. Thus SB is an over-sum effect. 'Over-sum' or 'over-summative' means a behaviour which results from an interdependence of component activities. Bertalanffy termed it 'non-summative' and 'non-trivial' [17:67f]. Here 'over-summative' is preferred, in order to relate to the parts / whole intuition of Aristotle: „The whole is something apart from the parts. It is more than their mere sum." [e.g. 3], commonly referred to as the 'over-sum principle'.

For an example take the ion channel of Figure 2.1D. Component behaviour is trivial: the random occupation and evacuation of a binding site by an ion. System behaviour SB is over-summative: a vectorial fluctuating net-flow of ions into or out of the cell. It cannot be predicted from component behaviour alone.

The over-sum effect is caused *exclusively* by the cyclic working assembly of components, effects not represented in the cycle-rate equation do not contribute. Clearly, SB can not be attributed to individual components alone, nor to a mere addition of component action. Rather, it must be attributed to the concerted action of components in the kinetic cycle of the STD.

The special kind of causation thus identified may be referred to as *'m-causation'*. Of course, it should and does match the pragmatic criteria of causation given above. The m-causation is characteristic for kinetic models of steady state mechanisms, as these necessarily have multiple transitions and at least one SB generated by a full cycle of states.

Importantly, as the term 'over-sum' suggests, a whole with its SB-property is 'more' than what the whole's individual component actions lead us to expect. In steady-state mechanisms this 'more' is made possible by the cyclic *arrangement* of components, the system structure stipulated in the STD. Yet, despite the new SB quality of being 'over-summative' or 'more than', the SB is reducible to (fully explained by) the interactive component--level.[26]

OSE: Over-sum effect.

Models of steady-state mechanisms are based on cyclic STDs. They cause 'over-summative' system behaviour, exceeding what individual component actions lead us to expect. Due to the over-sum effect of interactive components a whole with its SB-property can be 'more' than what the whole's components predict, but is still reducible to (fully explainable by) the interactive component-level.

---

26 In this case reduction is succesful and may be extended down to physical base. However, as will be explained, our comprehension may not follow, because of limitations of our cognitive abilities (PLC). This thread will be taken up in Section 4j and 5c.

3. Systems

In this chapter systems theory is critically discussed. The linkage of levels, be it causal or constitutive, is a difficulty. Systems of conventional theory are found to be systems of symbols. Any top-down influence from a whole to its parts is to be denied. 'Universal levels', which are linked by identity of their basal items, are introduced.

Systems theory attempts the quantitative modelling of parts of reality. Such models, though belonging to different scientific disciplines, may be structured in a uniform way. The ultimate goal of early approaches was their merging into an all-comprising hierarchical structure demonstrating the unity of science [e.g. 107].

3a. Levels

Phenomena of world and mind are ordered with a vertical stack of levels.[27] The levels are hierarchical in that entities of one level are directly dependent on several entities of the level be-

---

[27] The levels of description in physics, discussed above in Section 2b, provide an example. There, however, the guiding principle for the upward coupling of level-1 to -2 and -2 to -3 is not uniform, except for the neglect of information, which is common to both.

low, which relate to each other in a special way. If the guiding principle of level relations is that of *whole and parts*, as originally proposed by Arthur Koestler [81][28], then the whole at level n is constituted of its interrelating parts at level n-1, below. In turn, each of these parts is a whole constituted of its own parts at level n-2, and so on.

Given any two adjacent levels, systems theory attempts a predictive understanding of the behaviour of a whole, the explanandum, at the upper level from interaction of components at the level below. This 'reduction' is a recursive process (in my view) of quantitative *bottom-up* modelling.

Since parts (or components) are themselves composite objects (wholes), there is a downward cascade of wholes and interacting parts. When following this cascade, we presumably find the ultimate- or zero-level, where objects (elementary particles) have no parts. Laws are ultimately based on (may be reduced to) those of the zero-level. Whatever occurs at this level, interactions [29] or something else, cannot be reduced further, there is nothing below. Such downward 'drainage' is controversially discussed in [20, 73, 74].

Apart from whole and parts, several other guiding principles for building systems have been proposed. They were enumerated by C.F. Craver in his 'taxonomy of levels' [30]. In *levels of science*, particle-physics (with its language mathematics) is placed below chemistry, which is below biology. Here size comes in when the objects of the scientific fields are compared. Biology is below psychology, which is below sociology, etc. In *levels of nature* size is one of several guiding principles. The hierarchy extends from subatomic particles to microscopic objects, to macroscopic objects, to planetary systems, galaxies etc. Further guiding principles in systems of nature are causation (in processing or control) and composition (mereology, aggregativity, containment, constitution).

Well known is Oppenheim and Putnam's 1958 linear world

---

28 See also [17] as criticised by [15].
29 This as long as space and time, concepts required for interaction, remain applicable.

model with 6 levels of nature (or science): particles, atoms, molecules, cells, organisms and societies. Such levels are at the same time levels of size and are described in level-specific idioms. The possibility of reduction from one level to the level below assures the "unity of science" [107]. Wimsatt's world model is not linear but branches, accommodating more levels [140, 141]. Again levels are ordered by size and nature/science.

Also well known is Churchland and Sejnowski's concept of organization-levels in neuroscience [28]. Since neuroscience is part of science, the scheme may be viewed as a subdivision of 'levels in science'. It specifies levels of molecules, synapses, neurons, networks, maps, systems, brains (Figure 3.1). These levels of organization imply a hierarchy of increasing sizes and, accordingly, types of processing, where the processing on an upper level depends on the processing at lower levels.

Ordering of levels according to size, indirectly in the three examples mentioned, is compatible with the application of a parts-whole concept. This because a material part is necessarily smaller than its whole and the large is constituted of the small. Thus a parts-whole order is also an order by size.

It will be noted from Figure 3.1 that any item or event at a level also appears at the level above. To give an example: A synapse-*holon* [30] on level n-1 is implied to contribute to the integral change of membrane potential of a neuron-holon on level n. On level n the synaptic event of level n-1 is viewed in a wider context. It is the same event. Therefore linkage of size-levels is not by causal interaction (which may link different objects) but due to something else: say, by parts-whole 'implication' or by symbolic representation.

Indeed, in the theory of hierarchical systems the vertical coupling of system levels by causal interaction is disfavoured and the "downward causation" is a topic of much concern [e.g. 2]. According to a more recent suggestion by William Bechtel and Carl F. Craver, such causal interaction between whole and parts of a mechanism is a strained notion [10, 30, 31]. For how can

---

30 'Holon' is Arthur Koestler's term for a collection of interacting components at system level n-1, giving rise to an over-summative SB of a whole at level n, above [81].

the parts encounter their whole in space and time, as would be necessary for interaction, if they are contained in the whole? (See also John Danaher, who lists more reasons why causal linkage of levels is disfavoured [34].) Thus causal interaction should be restricted to intra-level events.

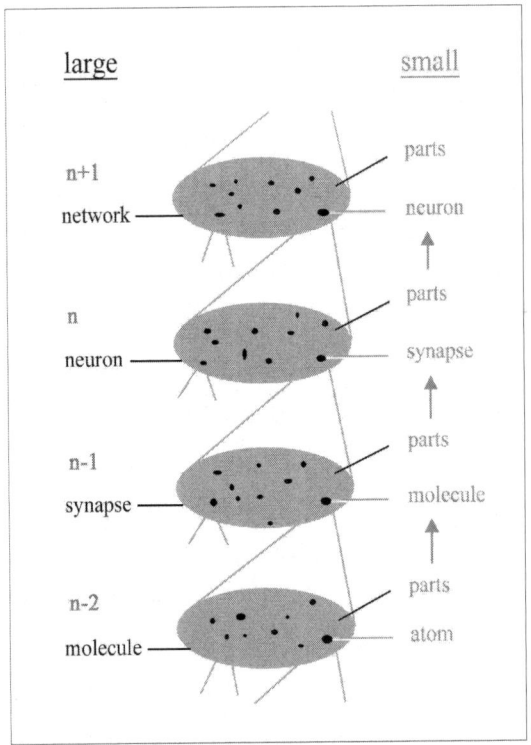

Figure 3.1. Some levels of organization, size and processing, as in the system by Churchland and Sejnowski [28]. Note that every item appears twice: As a 'holon' (in Koestler's wording [81], grey ovals, named on the left) and one level up as a *whole* representing this holon (black dot named on the right) in a wider context, zooming out. A large number of such holons reside in anyone system level (only one per level is shown). Every whole-item within every

holon of a level is a holon located on the level below.

3b. Linkage of levels by constitution

These considerations led to a version where system levels (local rather than global) were connected by *constitutive* dependencies of wholes of mechanisms [10, 30, 31]. The novel concept, however, carried the further notion that the constitutive relationship is symmetrical. This symmetry is not given, as will be shown.

Craver defines "constitutive" as a relationship "between the behaviour of a mechanism [31] as a whole and the organized activities of its individual components." [30:108].[32] Craver states further: "... all constitutive dependency relationships are bidirectional" [30:153], and (with William Bechtel) "The relation is symmetrical precisely because the mechanism as a whole is fully constituted by the organized activities of its parts" [31: 554].

Based on this apparent symmetry Carl Craver makes a strong point that *top-down* changes, directed from whole to components, are possible. They are initiated for instance by external agents interacting *top-top* with the mechanism as a whole and proceed downward synchronously (without delay) in a constitutive rather than causal way [30: 146,151,153]. This implies *mutual manipulability,* from components to whole and from whole to components.

Several authors give reasons to deny such mutual manipulabil-

---

31 In my wording, a mechanism consists of components in interdependent relations. They interact causally in an optimized way, generating events which jointly result in a distinct and more or less predictable system behaviour SB, which alters the environment.

32 It is claimed that the activity of the whole of a mechanism is constituted of the activities of individual components, as these work according to organization. It may be added that the activity of the whole amounts to more than a trivial combination of component activities. In such non-trivial cases the activity of the whole is constituted of the resulting *oversum* activity, here called SB. The latter is caused jointly by the interdependent component activities, as represented by the matrix of rate coefficients of the system of equations describing the mechanism [17].

ity. John Heil, discussing a text by Jerry Fodor, states as a common notion that "higher levels asymmetrically depend on lower levels" [60:29]. Peter Simons views asymmetry as the first formal property of a part-whole relation [125, 126]. Max Kistler maintains that the constitutive relation is asymmetric [75] and Petri Ylikoski also concludes that the relation is asymmetric and 'mutual manipulability' is not given [144].

Indeed, how can a whole be influenced *top-top* and how can it then change its parts *top-down* when it is exclusively dependent on these parts? Such exclusive dependence means that a whole can be changed only by changing a part.

Actually, a whole at level n can be changed only by changing its parts at level n-1, which can be changed only by changing their parts at level n-2, etc. The real change is at zero-level and all levels above show the same event at different scale (magnification, zoom). Thus, if system levels differ only by scale, there can be no inter-level effect. One magnification cannot affect another (see Section 3f).

3c. *Top-down* effects

Top-down is the direction from whole (upper level) to parts (lower level) in a conventional hierarchy of system levels.[33] Given a *bottom-up* constitutive linkage of levels, lack of symmetry means that the inversion, i.e. *top-down* linkage, is not possible. For if a whole is constituted of parts, it cannot affect these parts. Yet numerous authors propose *top-down* effects, claim pointedly that a whole can influence its parts. This dictum can be traced back to Seneca and other stoic writers.[34] It is considered well established even in recent literature [e.g. 2, 53, 54:178, 72:128, 127, 130, 131:249ff]. Is it wrong?

We have to distinguish general *top-down* effects from the causal *top-down* interaction (a special kind of *top-down* effect). While most of the above authors concede even causal *top-down* inter-

---

33 Not meant is the effect of a 'higher' brain module acting back on a lower stage of processing.
34 As quoted by F. Ast [4] and J.C. Smuts [127].

3. Systems

action, Bechtel and Craver convincingly deny it, arguing that a whole which contains its parts cannot encounter them in space and time, as required for causal interaction. Instead, Bechtel and Craver allow for constitutive *top-down* effects [30, 31]. But such inversion of the constitutive relation we have just excluded, the relation lacks the supposed symmetry.

Having excluded causal and inverse-causal as well as inverse-constitutive *top-down* effects, we must deny the possibility of *top-down* effects all-together (TDD). If parts constitute a whole exclusively, then the whole can be changed *bottom-up* only, by changing the parts. It is not possible to change the whole in a *top-top* way and then observe *top-down* effects on the parts.

TDD: *Top-down* denied.

For parts-whole systems the possibility of *top-down* changes is to be denied all-together. If parts constitute a whole *exclusively*, then the whole can be changed *bottom-up* only, by changing the parts. It is not possible to change the whole in a *top-top* way and then observe *top-down* effects on the parts. Any *top-down* influence from whole to parts, though an ancient holistic concept, is impossible.

In summary, any effect from whole to components, causally or inverse-constitutive, is full of pitfalls and contradictions, is impossible. Below I shall argue for an alternative, that levels are all-comprising, *universal*. They are linked by identity of their basal items and differ in the grouping of basal items and the assignment of symbols to the groups.

3d. Switching of levels

I have argued that the *top-down* path from whole to components is barred. It is not passable by causal *interaction*, for the whole cannot encounter for interaction in space and time what is contained in the whole itself [e.g. 30, 31]. It is not passable by *constitution* because the constitutive parts/whole relation is asymmetric. The whole is constituted of components, not *vice versa*. Little wonder, therefore, that indisputable experiments demonstrating *top-down* changes are not on record.[35]

Yet, in our common-sense understanding it remains indisputable that wholes can interact. *A stone thrown will smash a window as a stone,* irrespective of our academic knowledge that the stone consists of crystals and atoms. Here we simply *disregard* that the stone is constituted of components which cannot be affected by their whole. Instead we stress over-componential properties of the stone like size, shape and weight.

A herd of buffalo comes to the lake for drinking. Here we disregard that the herd is constituted of animals and that it is the animals which are drinking. By grammatical structure we pretend that the herd interacts with the lake in the act of drinking. *The troops march, sing a song, the armies clash,* we disregard the role of the soldiers, stress the over-componential properties of the army like numerical strength, tradition or nationality.

Apparently we can switch on and off the constitutive aspect of an object. When we switch this aspect off, we focus on the over-componential properties of the constitutive whole.[36] We disregard that even over-componential properties owned by a whole are directly dependent on properties of the components.

The pragmatic basic level : But let us feel free to select a level where interactions should take place. To avoid unnecessary detail we pick this level as high as possible, yet as low as necessary for reductive explanations. For instance, if we are interested in biological phenomena, we choose a sub-biological level of molecules as the interactive basic level. Then all our interact-

---

35 Certainly a convincing *correlation* is often found between phenomena on levels n and n-1, but this does not prove that a causal or constitutive *top-down* linkage from level n to n-1 was active for obtaining the correlation.

36 A constitutive whole is constituted of components, their relations and activities.

ive objects molecules. Above this pragmatic basic level will appear higher levels containing symbolic wholes.

Thus the level of interest is always made the pragmatic basal level. Here causal power is attributed to objects and lower levels are disregarded. Wholes at a level above the pragmatic basal level are without attributed causal power. They cannot interact, their role is prerogative only.

a) The macroscopic objects of the outside world, to which we conventionally attribute an independent existence, are in truth not independently existent. They are consequences of underlying meso- and microscopic objects and processes. There is a cascade of dependencies stretching downward from the macro- to the microcosm. This cascade is epistemic, is not independent from us, is due to our senses and our attempts to make order.

b) What is really outside, the zero level, is not accessible to our senses. Our sense organs appear to us as macroscopic objects and perceive stimuli seemingly emanating from macroscopic objects. These stimuli and objects are represented with our neuronal mechanisms and processed by them in various ways.

c) We represent the objects internally as wholes, *symbols* which arise from our neuronal activity of grouping, ranking and simulating. Epistemically there is a cascade of whole-parts drainage stretching downward to a pragmatically chosen zero level. Certainly wholes as well as system levels are not independent of our thinking but are its products.

3e. Systems of symbols

The various kinds of system levels may be pragmatic conveniences. Each is a model, justified if it serves its purpose and each is limited in its own way. Yet the models have much in common.

Being directly or indirectly size-ordered, like the models by Oppenheim-Putnam, Wimsatt and Churchland-Sejnowski, they may be translated into parts-whole models. This because the large is constituted of the small. Thus the small is somehow implied or **represented** in the large, as many small constitute a large. When we want detail, we zoom in on the small (level n-1), neglecting for the moment the overview. When we want the overview, we zoom out, surveying the large (level n), neglecting unwanted detail. The Koestler model is such a parts-whole model and the Bechtel-Craver model too, even though restricted to mechanisms and interacting parts.

What is meant by a 'whole'? Arthur Koestler described the whole as: "...something complete in itself which needs no further explanation" [81:48]. In George J. Klir's quotation of Jan C. Smuts the whole is "... a synthesis or unity of parts, so close that it affects the activities and interactions of those parts, impresses on them a special character, and makes them different from what they would have been in a combination devoid of such unity or synthesis" [76:41, 127].

In disagreement with Smuts' statement that a whole "... affects the activities and interactions of those parts", which describes an impossible *top-down* effect, I suggest to understand 'whole' at level n as a representation or prerogative, as a *symbol*. It refers to interacting parts and, where applicable, their over-summative action by m-causation.

Then the various types of systems have a common feature: in each case an upper level contains prerogatives, symbols representing groups of relata of the level below, and their interaction.

A whole is a symbol representing a holon.

The advantage of this my short-hand notation with symbols is that the levels serve for economy of thinking by abstracting from unwanted detail. For instance, the atom, the galaxy or the 'mechanism as a whole' are symbols of such denotates. Their apparent interaction at level n does not really occur there, but symbolizes interaction of groups of relata at level n-1. These, in turn, are symbols of relata at level n-2, etc. The levels are populated by prerogatives, their interaction is symbolic. The real interaction occurs at a level below, presumably at the zero-level.

I propose that the objects of a level are symbols "about" one or

3. Systems

several relata on the next lower level. The symbols refer to these relata and their interaction, they represent or imply them.

The systems of system theory are systems of symbols.

Note that the denotates of symbols are holons located on the level below. The holons may or may not 'mean' objects of the world. The system is a construct simulating the world with symbols, with prerogatives, introducing an order which the world may not have. System levels are tools for the ordering of representations, products of our thinking.

3f. Universal levels, linkage by identity

System levels (I suggest) are essential for epistemic reasons. They order world phenomena into palatable chunks, as our cognitive abilities are limited. For instance, our working memory is restricted to few items at once [96]. In the *universal* levels, here proposed, all basic events are physical per axiom (the RP-hypothesis). Changes on all levels are due to the same basic events occurring synchronic[37] on each level. The various levels contain the same basic items, ordered differently. And the principle of ordering may differ from level to level, as it does in the levels of physics in Section 2b.

Universal levels differ in the grouping and the assignment of symbols to the groups, in context of symbolic objects and events, and in the level-specific idiom. This is illustrated in Figure 3.2. Otherwise levels are *universal*, each is all-comprising. Vertical linkage, except by identity of basal items, is not needed.

Of course, the re-grouping must be grounded on nothing but known and documented lower-level phenomena, basal events and laws. Wishful thinking and fictive phenomena taken for real are excluded, will result in failure to reduce. Given this provision of *proper grounding*, the principle of universal levels implies reduction and explanation of any phenomenon, including immaterial phenomena, to lower-level events and, in the end, to physical base.

---

37 instantly, without delay.

Reduction: Since all objects above basal level appear as symbols, their interactions are symbolic too. True interaction occurs at the basal level only. Thus an explanandum is to be understood from a cluster of lower-level objects (also symbols) and their apparent interaction.

Successful reduction suggests consistency, that no mistakes were made[38] in grouping and assignment of symbols. When discussing reduction, above, from level to level a continuity of laws was assumed. This continuity can now be understood as a consequence of universal levels where the basic objects and events with their inherent laws remain the same from level to level (see Figure 3.2 and Figure 5.1).

There is an important *caveat* regarding reduction across multiple levels. Applied to but two neighbouring levels, reduction provides mechanistic explanation. However, applied to multiple levels it amounts to the skipping of intermediate explanations, hindering comprehension. The rule is, the more levels are included, the smaller is the explanatory satisfaction (PLC, see Section 5c).

USL: Principle of universal system levels.

The concept of universal levels is developed. The levels contain the same basic physical elements, each level is all-comprising. Thus basic events are the same, occur synchronic on each level. The level-specific objects are symbolic. They arise from different grouping of the basic elements and assignment of symbols to the groups.. Context and idiom are also level-specific. There is no need for a further coupling of levels at all. Coupling is already complete, each level being the same. Since basal events are physical by axiom (RP-hypothesis), the physical basis of all complex phenomena is implied.

---

38 Samuel Guttenplan states "Reduction is often a way of preserving the ontological security of some item." [56:536].

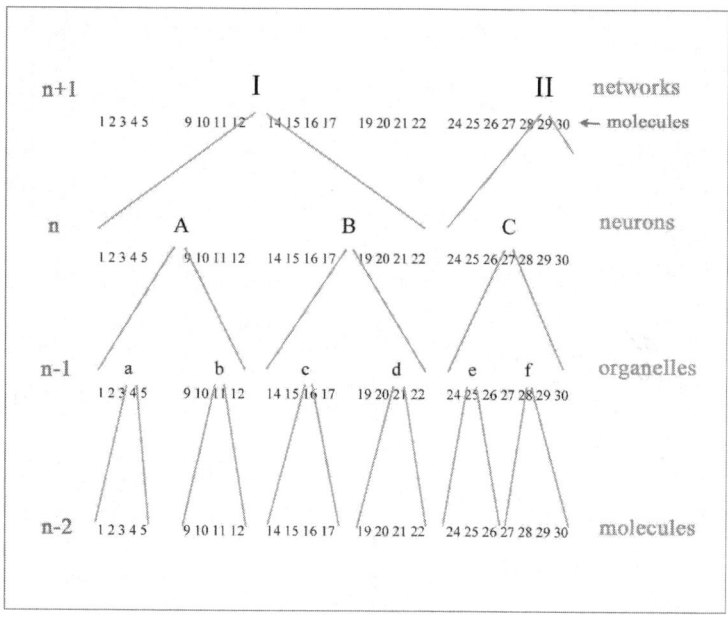

Figure 3.2. System levels of neuroscience showing level-specific ordering of the same basic phenomena 1, 2, 3... Group 12345 is symbolised as **a**. This, with **b**, is symbolised as **A**. This, with **B**, is symbolised as **I**. The network level n+1, with representations of neurons, implies molecules present in these networks and neurons (and events involving molecules) which are also represented or implied on levels down to level n-2. Thus the various levels contain the same, ordered differently. A symbolic interaction at a level implies equivalent synchronic events at all other levels. There is no need for additional vertical coupling of levels, their items already symbolize the basic items identical on all levels.

4. Reductive Modelling

Reduction is defined and several of its strategies are exemplified. The reduction of life, of agency, of art to physical base is explored. However, if feasible such multi-level reduction has little explanatory merit. Contrary to the literature, phenomena like qualia-feelings are functionally reducible, as they are found to be task-related after all. They do not provide an exception for RP. Post-reductionism is evaluated.

The term 'reduction' is used in the sense of explanatory reductions in science. Reduction is motivated by the "honest wish" to understand how things work [36]. It attempts to understand an explanandum from processes and their laws at lower system levels. Where we cannot reduce, we cannot understand. Our aim is to explore the feasibility of reducing explananda such as life, agency or mind to physical base. However, as a starter we shall reduce within physics itself, focussing on three levels of description detailed in Section 2b.

4a. Reduction in physics.

Referring to Section 2b, suppose we focus on level-3 and select one of the macroscopic laws as our explanandum. The law is a deterministic relationship, suggesting precise reproducibility.

We ask how the material constant specified in the law is related to other physical properties. The answer cannot be deduced from level-3. Rather, it is found by detailed research on level-2. Here measurements soon reveal that the material 'constant' is due to a process or mechanism with probabilistic behaviour. It causes variables to fluctuate from trial to trial and along time. Only when taking their mean values does the law hold.

Thus we have on level-2 found a process explaining the law on level-3 but obeying rules more general than the law on level-3. Further, we have found the special condition under which the law holds. The level-3-explanandum has been reduced to the model of a level-2-mechanism.

Now we face the more difficult task to understand the level-2-process in terms of level-1 phenomena. How can the level-2 fluctuations arise from deterministic processes on level-1? Research on level-1, when accounting for the full complexity of this level, is expected to result in very large models which predict the fluctuations deterministically. Then reduction to level-1 would be successful.

The procedure of reduction is this: In both the 3 → 2 and the 2 → 1 reduction modelling on the lower level yields a model behaviour which is to be compared with the explanandum. In a recursive process the model is changed until the match is satisfactory.

To exemplify this for a 2 → 1 reduction, take the growth of a particular snowflake as the explanandum on level-2. When observing its development repeatedly from identical seeds, one finds that the flakes are not alike under 'reproducible conditions'. Such fluctuations of structure are explained on level-2 by a chance process of apposition of water molecules to sites on the crystal. However, modelling on level-1 with full complexity replaces the chance process by deterministic causal chains. Recursive adjustment of model details finally yields a good match of model-flake and explanandum and identical flakes in each trial. The fluctuations can now be understood as resulting from insufficient attention to system complexity.

4. Reductive Modelling

4b. General strategies.

We distinguish *single-level* reduction, relating a level to one neighbour below, and *multi-level* reduction. The latter summarises a chain of one-level reductions, it reduces, say, psychological predicates (like 'remembering') stepwise to elementary physical events. As intermediate explanations are skipped, the result, though formally correct, may be difficult to comprehend. Therefore, subjectively convincing explanations are not expected from multi-level reductions. The scientific assessment, of course, relies on the formal proof, the explanatory satisfaction being secondary.

Regarding mechanisms, the task is to understand the behaviour and properties of mechanisms-as-wholes from the concerted action of their components at the system level below. This means to understand the cyclic component action as over-sum action (Section 2f). Reduction is possible when the empirical laws (or strong regularities) governing the explanandum do root in those of a lower level.[39] The laws will not be identical, due to different concepts and idioms describing the levels, but those of a higher level are explicable by those of the level below, may be reduced to them. In a way the known laws appear to have *bottom-up* validity, the seem to consistently extend to the level above.

Generally, successful reduction starts with a 3rd-person explanandum. When starting from a 1st-person explanandum and staying within the 1st-person perspective, STN as well as ANA and CCP (with mind immaterial) make reduction to neuronal base impossible. Fortunately a reported 1st-person phenomenon may itself be taken as a 3rd-person explanandum (starting point of Red B). This is indicated in Table 4.1, where the abbreviations are explained.

Some examples may be in order. The reported 1st-person experience "I have freedom of will" or "my immaterial thought causes action" would fall under "**Reduction impossible**". For within a subjective perspective a reduction to neuronal base is

---

39 In Nagel-reduction a theory A is reduced to B if all laws of A can be derived from laws of B [99], and thereby explained.

barred by STN, ANA and CCP. For varying reasons such impossibility was claimed in non-RP[40] and in post-reductionism.

Red A is the highway of scientific reductions, excluding the mental level. Reduction by model, by mechanism and by material synthesis belongs here (see below).

Red B nevertheless deals with the mental level, with 1st-person experiences, by changing perspective, treating such experiences, or their report, as 3rd-person explananda.

System level	Reduction impossible	Red A	Red B
n+1	Immaterial 1st-pers inner theatre stage experience		Reported 1st-pers stage experience, taken as a 3rd-pers explanandum
	↓ //		↓
n	// STN, ANA, CCP	3rd-pers explanandum	3rd-pers OSE
		↓	↓
n-1		3rd-pers OSE	3rd-pers OSE, but PLC
		↓	↓
n-2		3rd-pers OSE, but PLC	3rd-pers OSE, but PLC
		↓	↓

Table 4.1 **Synopsis of multi-level reductions RedA and RedB.** STN: subjective transparency of the neuronal, ANA: abstracta non agunt, CCP: causal closure of the physical, OSE: over-sum effect, PLC: principle of limited cognition. (Explanations summarised in Chapter 8.)

---

40 see Section 4f.

4. Reductive Modelling

One variety of Red B is *functional reduction*, where the 1st-person inner-stage experience is the experience of a functional sequence of events. This reported experience is taken as a 3rd-person explanandum. It is reduced by recursively seeking that neuronal process (3rd-person), which is *mentally experienced* (1st-person) as the specified causal role, task or function (see Section 4h).

Turning to the basic reduction *Red A:* how is it performed? With the explanandum at system level n, we cannot *deduce* the laws at level n-1 from those at level n. This is not possible because the idiom is different and detail is lost when assuming the point of view of an upper level. Rather, the laws at the lower level are found from independent research at this level, n-1. Then, to reduce a system behaviour at level n to level n-1 we make use of a recursive *bottom-up* procedure of modelling.

Let us distinguish *reduction to a model* and the less speculative *reduction to a known mechanism*.

4c. Reduction to a model (Red A)

We construct a working-model at level n-1 (at the 'bottom') from hypothetical components and their relations, using known phenomena, properties and laws of the lower level. We note upwards the level-n-behaviour SB* of this model and compare it with the explanandum SB. In a recursive process we modify the model (educated guessing, abduction) until its *bottom-up* behaviour SB* agrees with the explanandum. We may thus find the most satisfactory model for the SB-data at hand. Then we have reduced SB, explaining it with interactions of a best-fit model at the next lower level. The possibility cannot be denied, however, that another model, allowing an even better fit, may be developed in the future.

4d. Reduction to a mechanism (Red A)

We do research at the micro-level (n-1), searching for candidates among established (models of) real mechanisms. We determine their SB* and compare it to the explanandum SB at the macro-level n. In a recursive trial and error process (educated guessing, abduction), we search for that mechanism, which matches all known features of the explanandum, such that SB* = SB. In experiments, a modification of the mechanism should modify SB and an inhibition of the mechanism should abolish SB. If such *bottom-up* manipulability holds, the mechanism will be our reduction base. (In case of multiple realisation more than one mechanism may show a good fit. Then the data base must be expanded by further experimentation until one realisation drops out.)

Both strategies are common scientific practice. To give an example for reduction to a known mechanism, suppose there is a **pacemaker neuron** which generates spikes in a seemingly random sequence.[41] The neuron is not driven by other neurons but fires 'spontaneously'. Thus, on system level n of neurons (whole) and spikes (behaviour of the whole) we find output without input: it would seem that a causal chain begins here. The explanandum is effect without cause (objective chance), 'Erstverursachung', suspected to be impossible.

To understand this neuronal random generator we do research on system level n-1, the level below. We find sets of fluctuating ion channels, mechanisms which we can model probabilistically, specifying cyclic STDs. The current fluctuations result in a noisy membrane potential near spike threshold: an irregular spike train results. Several cycles of modelling are needed to get the details right. If finally agreement of model behaviour and explanandum is satisfactory, we have reduced the 'effect without cause' at level n to a causal chain at the level below.

Another example is the explanation of **active ion-transport**. In living tissues one observes an accumulation of potassium ions

---

41 The output of single cortical neurons is often surprisingly irregular, it has a random component or is completely random [e.g. 29, 78:350ff, 80:313]. The randomness means that redundancy has been removed and code optimized for limited bandwidth [78:373]. Of course, the population average of many of such neurons acting in parallel amounts to a more reproducible response.

in cellular compartments. This accumulation occurs against a pronounced electrochemical K^+-gradient. It amounts to a decrease in entropy, though in an open system, locally. That was the explanandum.

Disregarding fuzzy explanations like "life forces", research at the level below (molecular level) has discovered distinct protein molecules present in cell membranes, the ion pumps. Some of them convert free energy of ATP (a metabolite) into uphill movement of K^+. To get the details right, the pump-mechanism had to explain that sodium ions are moved out of the cellular compartment as K^+-ions are moved in. Indeed, the stoichiometry of the pump was determined to 3 Na^+ / 2 K^+ / 1 ATP for every cycle of the pumps STD. This was also the stoichiometry of the macro-phenomenon. Further, the macro-phenomenon was inhibited by a ouabain (a glycoside), which turned out to be a powerful inhibitor of the Na^+-K^+-pump molecule. Also, the voltage dependence of ion pumping was accounted for.

In the end, for every known feature of the explanandum the Na^+-K^+-pump and its regulation provided the explanation. Thus the macro-phenomenon "accumulation of K^+-ions" at level n was reduced to a Na^+-K^+-pump mechanism at level n-1. Gradually the pump-hypothesis changed to certain knowledge, guided by reductive modelling.

Of course, in the future additional features may be discovered which make a distinction between several types of Na^+-K^+-pumps necessary. (This topic, related to multiple realisation, will be touched upon in Section 4i.)

4e. Reduction by material synthesis (Red A)

In the spirit of "What I cannot create, I do not understand"[42] there is a tradition in biochemistry to assemble a complex system from its components and modify these until the overall performance matches that of the natural system. The chemical synthesis of the simple biomolecule urea was an early beginning.

---

42 Attributed to Richard P. Feynman.

The recent synthesis of genes, using natural enzymes in the test tube, is a more spectacular example. Further examples are many.

For instance, the complex system of glycolytic enzymes was studied *in vitro*. Thereby the intrinsic oscillations of the assembly were discovered and analysed [62]. A cell membrane, planar or vesicular, was assembled as an artificial bilayer of phospholipids. Purified membrane proteins where then inserted into the bilayer and their function observed [98]. Here the visionary aim was to synthesize a living cell and to demonstrate its agency.

Where systems become too complex for *in vitro* analysis, research turns to mathematical models for simulated synthesis. Integrative physiology tries to convert the usual "box-and-arrow" diagrams, which sketch out causal chains, into a quantitative formalism of differential equations. The goal is to model large parts of living organisms, e.g. their energy metabolism or their hormonal systems or their respiratory and blood circulatory system, as a quantitative interplay of the various organs and molecular components involved [50].

In any case such reductive modelling, be it by material or simulative synthesis, requires recursive *bottom-up* procedures.

4f. Life physical or more than physics?

Is life physical? Is the mind, part of human life, physical? RP answers with a resounding "yes", but there are other voices [e.g. 66]. The debate about life calls upon the study of biological objects like organisms, cells and protein molecules. To recall a few positions, reductive physicalism maintains that there is *nothing* beyond the physical and what seems beyond, including biological objects (and the mind), can be or will eventually be reduced to physical phenomena [e.g. 94]. Non-reductive physicalism agrees that everything is physical, but claims [e.g. 12] that some items are not necessarily reducible (see Table 4.1). Post-reductionism holds that the principle of reduction is doubtful as physical reduction base is incomplete and reduction to physics

4. Reductive Modelling

cannot provide the full explanation particularly of mental phenomena[43] [e.g. 69:201,202]. Apparently the mind of higher organisms, while depending on the neuronal (physics), is "more than physics".[44]

Here I intend to show *by discussing reduction* that living objects or agents generally are "more than" physical objects, and to explore the origin of this "more than". I find that they are "more than" but still reducible.

Explanatory reduction starts with a question, an explanandum. This may be the *behaviour* of an agent, which reduces to that of a biological object, e.g. a cell and its mechanisms. A subsequent explanandum may be the *existence* of this object. The existence is explained by the objects *genesis*. Research has shown that genesis (ontogenesis) is based on a *plan* consisting of genome and epigenome.[45] Together they contain the information describing the 'idea' of the functional product and also the information of ontogenesis, how the biological object builds itself by protein synthesis and autopoiesis, exercising its agency.

Based on the DNA sequences of the genome, autopoietic grows would lead up to the genotype of the organism. However, what results from such growth is not the genotype but, due to gene methylation and other epigenetic effects, something more advanced, the phenotype. Thus the plan consists of genome and epigenome, both are indispensable. A variety of such plan is owned by all biological objects, which may be classified as *planned objects with agency* without exception.

---

43 For instance, humans have agency, intention, values, dignity, properties which are hardly concepts of physics. Yet, can they be reduced to physics?

44 My wording. "M more than P" means that M depends on P, yet something contributes to M which seems irreducible to P.

45 Genetics is concerned with genes which specify, by a sequence of DNA symbols (the code), the sequence of amino acids in the first version of a protein. Genetics further deals with gene-controlling sequences. Epigenetics is concerned with subsequent changes leading to a "stably heritable phenotype resulting from changes in a chromosome without alterations in the DNA sequence". See http://en.wikipedia.org/wiki/Epigenetics and [e.g. 136].

What is a plan? A script describing with symbols events to be realized in the future. An *immaterial plan* (like a blueprint or a symphony score) consists of symbols for concepts (abstracts) or events and, being immaterial, cannot act causally by itself. A *material plan* is a material object somehow referring to the future genesis of a planned object. The plan is a script of material patterns decodable into information. Before realization of the plan the patterns are symbols describing the genesis of a planned object but not yet effecting this genesis. Later, during realization, the patterns are direct causes of events of genesis, or are causally relevant for the genesis.

This is in contrast to general *physical objects* like atoms and subatomic particles, but also supra-atomic physical systems like crystals, oceans, galaxies. Supra-atomic objects can be located in space and time. They are material objects consisting of smaller material objects which interact according to laws specifying physical forces and their effects. Every interaction happens by them. There may be a certain determination arising from such forces and laws, but there is no pre-existing plan. Consequently, the term "plan" is absent from physics textbooks. For real physical objects are not symbolic, are not "about", and they have no agency and no concept of the future.

Biological objects or agents resemble physical objects in that they can be located in space and time (are not abstract) and events within them, like those outside of them, follow chains of physical interactions. Thus the living objects are part of the physical world. However, there is a difference: The existence and ontogenesis of biological objects is based on a material plan, the genome and epigenome. These play a leading role in biology textbooks. In contrast, physical objects are unplanned. Its plan and agency makes a living object "more than" physical.

What about the genesis of the material plan itself? The genome underwent *evolution* which is based on a few types of arguably physical events. There is selection of individuals driven by environmental pressure, and stochastic variation of genes. These processes are unplanned causal events. Evolution, which explains existence and behaviour of genome and epigenome, is spontaneous, unplanned. Thus the material plan arose from unplanned physical events.

4. Reductive Modelling

Reduction explains a *phenomenon* at system level n with a mechanism or a set of micro-properties at level n-1, below. The phenomenon (explanandum) may be a *behaviour*. Referring to an example already mentioned, the transport of solutes against a concentration gradient once was a much discussed explanandum. Research at level n-1 showed it to be effected by a class of membrane proteins (pumps) consuming energy-rich phosphate bonds. This research allowed explanatory reduction from behaviour to a mechanism which was operating exclusively with physical forces. Such reduction, though a gold standard in the biosciences, does not yet address the mechanism's genesis.

Now suppose a subsequent explanandum were the *existence* of the molecular pump. Its reduction would lead to a material plan of protein synthesis as the reduction base. Further, taking the existence of this plan as explanandum will lead to evolution as reduction base. Evolution, in turn, reduces to physics, because its mechanisms are physical, even though its objects are material plans.

Here are examples of reduction to a material plan:

Protein molecules: Apart from some simple peptides synthesized at the surface of cosmic dust, protein molecules are of biological origin, synthesised by cells. The first version of a particular protein molecule is produced according to a gene found in a material library (genome) unique for the species. The gene and its genome underwent evolution and its existence is explained by evolution. The evolution, in turn, happened spontaneously, without a plan, and is explained by a few (known) rules based on causes and effects, operating on material plans.

A gene contains a sequence of patterns (letters, symbols) specifying a sequence of amino acids in the first version of the protein. The sequence of letters constitutes part of the material plan, it is information which exists prior to realization of the plan. But during realization the symbols become causally relevant, guiding ordered protein synthesis. By physical interaction (of being scanned) they determine the amino acid to be inserted into the string generated.

With the explanandum being a *behaviour*, explanatory reduction would reveal the mechanism of the finished protein, the functional phenotype (not of the genotype or first version). With the subsequent explanandum being the *existence* of the protein, explanatory reduction would reveal, and in fact research did reveal, genesis and the material plan (genome plus epigenome). Implicitly it would also predict the general behaviour (will it bind glutamate?) of the finished product.

Organisms: A living organism is an agent consisting to a large part of molecules. Water and other simple molecules aside, they are of biological origin, synthesised by cells. The molecules constitute polymers, cells, tissues and the organism, the genesis of which depends on an assembly plan unique for the species of organism. This plan, genome plus epigenome, is a *material plan*. Genes exist prior to plan-realization as scripts of patterns describing future events, in this phase the patterns are symbols (or we tend to interpret them as such). However, in the phase of realization the genes, being material, act causally or are causally relevant, effecting ontogenesis of the organism. This reduction of existence to genesis is not speculation, of course, but the confirmed result of experimental research [e.g. 136].

In summary, it can be shown by reduction of existence to genesis that biological objects (agents) are all planned. This makes them different from physical objects, which are unplanned. And it makes them "more than" physical because plans generally are not part of physics. Thus there is something substantial beyond the physical, the abundance of planned living objects argues against "nothing-but"-physicalism, or so it seems.

Going beyond: However, what do we mean by "M more than P", like in "Music is more than vibrations" or "A plan is more than physics"? That M is related to P in a special way: It is based on P but somehow goes beyond. A conceptual connection between M and P may be hard to find, for something contributes to M which is or seems irreducible to P.

This "something", I suggest, stems from the human mind. For the mind alone has a concept of "future", it alone can generate or read a "plan" as a script about the future. This is obvious in case of "immaterial plans" as the mind is the domain of the immaterial. It is also true for "*material plans*". These, prior to realisation, appear to refer with symbols to future events, where

4. Reductive Modelling

"symbols" and "future" point to a mental origin of "plan". In fact the genome is just a material pattern capable of interaction when circumstances allow. *Humans attribute to it the features of a plan.* Little wonder that the mechanisms relating to a material plan, or to its evolution, are all physical.

Thus living objects are objects based on physics and reducible to physics (see below), to which humans attribute the label "planned". This label (and others) makes them appear more than physics while (objectively) they are reducible to physics.[46]

> My point:
> "More than physics" is compatible with reduction to physical base.

4g. Is reduction of life feasible? (Red A)

Let us turn from living objects to life itself. According to the promise of reductive physicalism it should be possible to reduce even a complex process like 'life' to the interaction of physical components without remainder. However, such a multi-level reduction is a task which many will deem impossible. There may even arise a feeling that such an attempt transgresses the border of the permitted. Focussing on the difficulties, we note that life is traditionally described as constituted of a number of sub-processes. Indeed, life may be the *over-sum* system behaviour [47] arising from a bundle of sub-processes. Then the reduction of life can be attempted by first reducing to sub-processes, then re-

---

46 This matches our observation of Section 2f that the over-sum-SB of a physical mechanism is "more than" what is expected from the components alone, yet clearly reducible to physical base.
47 'Over-sum' as in m-causation (Section 2f). Compare Dennett's *weak emergence* in the "innocent" sense [38].

ducing the sub-processes separately, then dealing with a possible remainder.

The sub-processes are (my disposition): Evolution of organisations which assure the realisation of a large number of distinct causal chains or mechanisms, including explicitly (1) metabolism, (2) homoeostasis, (3) response to stimuli, (4) communication, (5) growth. These complex mechanisms depend on each other, communication requires metabolism, reproduction requires response to stimuli, etc. Further: (6) Agency, (7) inheritance of genetic information based on a material plan, (8) reproduction, (9) autopoiesis, (10) stochastic variations of the genome (mutability), allowing for adaptation of the population to a changing environment by selection of individuals with advantageous mutations.

The above bundle of interdependent sub-processes is realized by all cellular and multicellular living organisms. The bundle is not displayed by lifeless physical objects like galaxies or crystals. However, the first five items are *individually* realised in man-made non-living objects, machines, automata. Since these are made of physical parts, there is little doubt that the first five items are separately reducible to physical mechanisms by material synthesis. Items 6 -10 appear to be more life-specific. Yet in principle their reduction seems possible, too. An example is

Reduction of agency: An agent is an entity which has autonomy, it makes decisions and acts upon them. Among living entities the smallest (arguably smallest) autonomous agent, or partially autonomous agent, is each of the 100 million living cells of a human body. Based on cells, in multicellular organisms a hierarchy of agents is found: An organ is an agent constituted of cellular agents; the organism is an agent constituted of organs and other function-realising systems. The hierarchy culminates in conscious agency, the first-person Self.

Can agency, can autonomy of decision be reduced to physical mechanisms? Well, decision processes abound in technical systems. A refrigerator is perhaps the most-quoted realisation. Autopilots have been constructed and are commonly used in ships and aircraft. Software-agents are an important device of

4. Reductive Modelling

modern computer programming. Thus at least "we can make it" from physical components, *reduction by material synthesis* assures us that agency has physical roots. Further, in neurophysiology feedback circuits well understood (like regulators of core temperature, of breathing, of blood pressure, retinal light-adaptation etc.) demonstrate successful reduction of decide-and-act agency to biophysical mechanisms.

On the whole, then, there is no principal difficulty in reducing cellular and supra-cellular agency to interaction of physical components by material synthesis or simulated (computer-) synthesis. The agency of whole organisms with their agent-world polarity still awaits such synthesis. Meanwhile its reduction may be attempted using the above reduction-to-a-mechanism strategy. The hierarchy of agency provides a staircase of reduction which may be followed downward.

There is one peculiar feature which machines as well as living beings possess but non-living physical objects, if not artefacts, do not. This, as detailed above, is their overall organisation, their *'idea'*, their construction plan. A technical idea, perhaps taking the form of a STD, originates in the mind of, say, an engineer. Thus the idea of a machine depends on the existence of its human constructor. For living beings the idea or plan of individual organisms seems to have arisen from modest beginnings by an evolution of species. The idea is contained in the genome / epigenome and amounts to an assembly plan of the genotype and its modification to the phenotype, to be realized by autopoietic growth.

The remainder: We have reached an interesting turn. While we may reduce the various sub-processes of life, including agency, to physics, it remains to consider the overall *idea* of a living organism, the *building-plan* contained in its genome. Again we face the challenge to understand what an idea or plan really is and to see whether it, too, can be reduced to physical processes.

Here is a summary, slightly simplifying the situation: Living organisms are constituted of molecules. A reduction of molecules to physical parts is possible, it is standard scientific procedure. The molecules are arranged into supra-molecular structures seemingly in accord with an assembly plan. This plan appears to be contained in the genome and allows for autopoietic grows leading up to the genotype. What results from this growth is not the genotype, however, but, due to interaction with the environment and other effects (epigenome), something similar but more flexible, the phenotype.

What is a plan? As discussed above, it is a script 'about' the future, describing with symbols events to be realized at a later time. The autopoietic assembly of molecules and supra-molecular structures like cells appears to realise such a plan.

Note, however, that concepts such as 'future' and 'idea' or 'plan' are abstracts. As such they are products of our mind/brain and not independently real. Thus they cannot be 'contained' in the material genome. Contained in the genome are molecules which will react in a certain way when conditions permit. The reaction changes conditions and thus allows further reactions along an optimized causal chain. To describe this chain or network of reactions as due to a material plan may be justified, but it is *our* plan or idea, a product of the analysing mind. In physics there is no pre-existing plan. Real physical objects are not symbolic, are not 'about', and concepts like 'future', which are not independently real, are not 'known' to them. The mind alone has a concept of the future and can attempt to predict future events.

First is a material genome. Then, if conditions permit, happens what we describe as realisation by autopoiesis. Any idea or plan, seemingly preceding the realisation, is in fact retro-engineered from the realisation - by our mind/brain. Therefore, in the evolved organism like in the designed machine, it is the human mind which creates or associates a plan. Reduction of a plan will depend on reduction of the mind (see Chapter 6). A plan in the material world, independent from our thinking, does not exist. There is no remainder.

In summary, life is viewed as the over-sum system behaviour based on a bundle of interdependent sub-processes. Reduction of life to its sub-processes, using the mechanism-strategy, seems feasible. Further, the sub-processes may be separately reduced to physical base using for instance the material-synthesis-strategy. Involving many steps, this multi-level reduction (once it is performed in a formally correct way) may be unconvincing. Then the impression will remain that life is more than physics, with agency and autopoiesis it seems on a higher organisational level. Indeed, in a way this is the case: Life as over-sum behaviour is more than the sub-processes and their relations. Yet, like other over-sum effects, it is reducible to physical base without remainder.

Its successful reduction to physical base (even though of little explanatory value) in no way belittles the splendour of life's evolutionary course, culminating in the refinement of conscious agency. It need not lessen the wonder which humans feel when unravelling, in an ongoing effort, threads and texture of the life process. Regarding vitalism versus physicalism, the miracle-feeling of vitalism remains while the attribution of the miracle to a super-natural power is replaced by attribution to 'nature'.

4h. Functional reduction (Red B)

Specifically for the mind-neuron relation Jaegwon Kim proposed a general strategy of reduction by *function* [74]. With the mind being not immaterial in Kim's view, the strategy relies on an identity of functional causal roles at the two system levels. In my words:

> (a) The functional causal role of a mental macro-property is identified.
> (b) A neuronal micro-property is identified which has the same causal role. This may be a process with many interacting components.
> (c) The two properties are taken to be identical.

Here 'functional causal role' refers to the causing of a behaviour in a functional context. In a) a macro-property (macro-behaviour) SB is caused in an initially unexplained way. In b) a micro-property SB* is caused by neuronal mechanisms. If behaviour of the same causal role (i.e. the same behaviour SB = SB* in the same functional context) results on mental and neuronal level, then reduction was successful, mental SB being explained by the neuronal mechanism causing SB*.

For Kim the mental is not immaterial. Therefore, Kim can attribute causal role and function to mental processes. Then he can compare or equal mental to neuronal causal role, task and function.

In our understanding, for good reasons, the mental is immaterial.[48] Then it cannot have a causal role or function (which requires a material process). Rather, the mental is an *experience* of the causal role and function presumably of a material neuronal process (the process itself being transparent). Then functional reduction means to use Red B, to seek the neuronal process (3rd-person) which is *mentally experienced* (1st-person) as the specified causal role, task or function. Further, it is expected that *bottom-up manipulability* holds, i.e. interruption of the neuronal process will abolish the mental experience.

Given this outcome there is no principal difficulty, the functional approach can yield a defensible hypothesis reducing the mental explanandum to a neuronal mechanism. Of course, when applied to a particular case, this will require much detailed neuro-mechanistic experimentation and recursive *bottom-up* modelling.

4i. Qualia-feelings

In reduction multiple realisation may be a problem. A mechanism producing the explanandum may be one of several and not the one realised. Further, in both of the above strategies of functional reduction any macro-properties without functional role (without SB) cannot be reduced. But is this a problem? Are there mental properties without (experienced as well as objective) functional significance?

Kim argues that the much-discussed *qualia*, the subjective quality properties of our raw-feeling experiences, are of this kind. Qualia, such as arising from tasting a wine or having a headache or sensing a colour, Kim identifies as not task-oriented, to him

---

48 See below, Section 6h.

4. Reductive Modelling

they are feeling-properties without function. Thus they constitute the irreducible residuum of an otherwise functionally reducible mental world.[49]

However, it is not certain that qualia, and feelings generally, are without task or function. Feelings are evaluated when the present sensory scene is associated with the present state of body-sense. Thereby the scene obtains a value indicating bodily well-being. I suggest that, as 'raw feels', qualia classify as feelings and evaluate to, say, pleasant, neutral or dislikeable, even if the detailed quality of the 'raw feel' cannot be verbalised.

Thus qualia are task-oriented in the sense of Damasio's markers of body states. By their temporal association with synchronic outer-sense percepts the markers help to avoid the unpleasant, the unbecoming scene and seek the pleasant, becoming scene.[50] Antonio Damasio pointedly described the functional role of evaluated feelings as guides of our behaviour [33:55]. Considering this, qualia should be classified as functional. By their temporal association with multimodal scenes they evaluate these scenes. Then the hypothesis of a general reductive physicalism extends to qualia issues.

4j. Post-Reductionism

When looking at the often irrational way scientific knowledge is produced [77, 83], it is not surprising that the resulting picture of the world is 'dappled' [24, 46, 69]. For instance, quantum theory (about the very small) and the theory of relativity (about the very large) are not compatible. There are spotlights but not enough overall illumination. As Steven Horst remarked, due to limitations in our cognitive faculties our scientific models are

---

49 It is remarkable that not phenomena such as design, agency or dignity are considered irreducible, but raw feelings. See reduction of agency in Section 4g and of dignity in Section 6g.
50 Usually it is concluded that qualia have no Brentano-intentionality, are not 'about'. However, they are associated by synchrony with behavioural context, with multimodal outer-sense percepts. So, in a way, they report about this context.

local, piecemeal and idealisations. There are explanatory gaps not only at the mind-body junction but at many junctions [69:5]. The result is not the intended unity of science but cognitive pluralism with a frequent lack of reducibility.

This deficiency gave rise to a post-reductionist stance were a successful reduction is an exception rather than the rule. Moreover, naturalism and reductionism cannot possibly achieve the complete reductive explanation of mental phenomena. For this failure S. Horst gives the following reasons [69:201,202]:

> (a) "Not everything about the mind can be understood in non-mental terms. We are natural beings, *among other things*. Some of those other things cannot be explained by appeal to the natural sciences." In consequence reductionism is false and harmful. It threatens our core identity as human persons and free agents.
>
> (b) There is no unitary, all-compassing model of nature. Thus a reliable reduction base does not exist.

In my view, argument (b) is acceptable with the addition "but they are working on the problem." For there is the hope that such a unitary model will be approached by and by. Meanwhile, the present model is demonstrably complete enough to serve as base for some reductions. Where we cannot reduce, we cannot understand.

(a) Indeed, when limited to the 1st-person perspective, reduction of mental to neuronal phenomena is impossible (Section 4b). Fortunately there is the alternative Red B. Disregarding it, post-reductionists fall back onto their 'more than a machine' intuition. They even see RP as a threat to our identity. But the real threat is an illusionary identity. This must be corrected, and be it by reduction.

Note that an alternative either 'more' than physics or reducible to physics is questionable. Due to the over-sum effect, something like the SB of a whole can be 'more' than the whole's assembly of physical components but fully explainable by component interaction. It is reducible to component-level and, in the end, to physical base. However, I concede that, if the mental were reducible to physical base, this indirect, multi-level reduction would be of little explanatory value. Thus argument (a) correctly restricts itself to 'understanding' and 'explanation' (presumably within our cognitive limitations). The case will be illustrated with an example below.

We distinguish *single-level* and *multi-level* reduction. Let us focus on the multi-level variety, possibly leading to a unity of science. Faced with the diversity of scientific disciplines, with idioms quite remote from the idiom of physics, a possible link to physics must be very indirect. A subjectively *comprehensible* unity of science may not be achievable by such multi-level reduction. A unity of explanations certainly is not attainable, for the more of system levels we include in reduction, the weaker is the explanatory return, the 1st-person aha-experience of comprehension. This is because intermediate level-to-level explanations are skipped, making multi-level reduction subjectively unconvincing. Best comprehension is obtained by one-level reduction. The reason suspected is indeed epistemic: The processing of our brain is limited in many ways (PLC). For instance, our working memory may contain only few items at once [96].

Therefore, where we need reductive *explanations* (and we need them), we better focus on one-level reductions to preserve the subjective explanatory value.

Reducing art? Let us probe the situation with another example. Consider a work of art, like a statue of Apollo. An ancient artist has hewn it from a block of marble. How to reduce the Apollo statue and its aesthetic value to physics?

(a) Its bulk consists of marble. This is easily reduced to physical atoms.

(b) Its surface follows a peculiar spatial pattern which has emotional appeal to humans. The pattern simulates the surface of a human-shaped deity, one of exceptional beauty. The statue is a message from the artist to the beholder about the deity.

To reduce this aesthetic *message* to physics by multi-level reduction may be possible, though it is indirect and tedious. Since intermediate explanations are skipped, multi-level reduction carries little explanatory value. This lack, which is due to third-person cognitive limitations, may give rise to the notion of 'art being more than physics'. In a way this notion, with 'more than' perhaps due to an over-sum effect, is justified.

Better explanatory value is obtained by one-level reduction, for instance from the Apollo statue to a general theory of aesthetics or to a general theory of communication. Of course, despite any reductive explanation the Apollo statue will keep its emotional appeal and remain beautiful art in its own right.

Similarly, a human, say, may be objectively reduced to physics but the inherent explanation is not subjectively *convincing*. People may have the first-person experience of being a dignified person and free agent. Yet a multi-level reduction of this experience to physical base, if successful, will necessarily be unconvincing. It will tend to support the notion of being "more than...".

In fact, however, a human agent as a whole may be at once 'more' than its reduction base (first-person experience and third-person over-sum effect), not explained by physics (first- and third-person cognitive limitations) and fully reducible (third-person multi-level reduction to physical base). Despite this reduction the first-person experience of being a dignified person and free agent, of course, remains.

In conclusion, the arguments of post-reductionism are justified in part, but overextended. They do not at all support a general refutation of reductions. To give up explanatory reduction means to terminate our strive for explanations, for understanding.

5. Neuronal Systems

Neuronal systems are systems of mechanisms working in parallel as well as in series. Neuronal communication is being unravelled by intense research. The code appears to be complete, contrary to the claims of constructivism the contextual part is not missing. Traditionally the brain is described with a hierarchy of organisational levels ranging from molecules to neuronal circuits and beyond. The vertical linkage of levels (causal? constitutive?) is problematic. Universal levels with their linkage by identity of basal items offers an alternative. The principle of limited cognition (PLC) is introduced and exemplified. The possible merits of random fluctuations are noted.

Neurons are the cardinal cellular agents of our nervous system. They are built of elements with stochastic behaviour, molecules. Neurons receive multiple inputs from synapses and generate output at multiple locations, again with synapses. These are one-way communication devices. The input synapses are connected to the neuronal cell body (soma) by trees of thin protrusions (dendrites). The soma connects to the output synapses by a nerve fibre, the axon, conducting nerve pulses. Due to a special method of 'saltatory conduction' the axons can be meters long but thin, and still conduct pulses with speeds of up to 140 meters per second.

The system of neurons is autopoietic, it unfolds and builds itself from modest beginnings according to inherent rules. Guided by molecular markers at the cell surface, multiple synapses are established at selected target neurons. The final complexity is impressive: Our adult nervous system has about a thousand trillion synapses linking about 86 billion neurons [79:16]. In one cubic millimetre of adult cortex alone there are about 3 km of thin axons serving for neuronal communication [124:470]. The code of this communication, which varies with the stages of signal processing, is under intense investigation.

5a. Neuronal code and the missing context

One-way communication is the presentation of a physical pattern by a sender, triggering a specific reaction in a receiver. The pattern may be a molecule (e.g. a hormone), a sequence of sound waves, a series of marks on paper. The receiver would be a neuron with a hormone receptor, a bat, a human reader. The receiver interprets the pattern as signs denoting a meaning. The meaning is the propositional aspect (content) of information. Originally meaning was neglected in Claude Shannon's famous mathematical theory of communication [122], and there is still no theory covering it.

The 'decoding' from pattern to meaning requires that the receiver is in possession of a code table, for instance by previous learning, which lists the signs and allows their association with meaning. In "transmission of information" only the physical pattern travels through space and time. The abilities of interpretation and association remain faculties of the inaugurated receiver. Thus the meaning is not transmitted but was learned previously and may now be reconstructed from memory.[51]

Data: In neuronal communication sender and receiver may be sensory cells, neurons, muscle or gland cells. They present and respond to patterns which stand for data. Generally, data are structured as follows:

---

51 Information, in a common phrasing, is the 'removal of uncertainty'. Such removal cannot be transmitted, but is enabled by transmission (of a pattern) acting as a trigger.

1) There is a variable part, a quantitative chunk. This may be a number or a continuous variable answering the question *how much?*

2) There is a fixed part, a contextual legend. This may be a series of symbols denoting what the quantitative chunk is about. It provides the meaning of the *how much*, answers questions about the *what, where* and *when*.

In neuronal communication only the *how-much*-information is represented by a pattern of nerve pulses (more precisely by a pattern of time intervals separated by nerve pulses). In the range milliseconds to seconds the intervals are suitable for mapping *analogue* quantities, and their evolution, by analogue symbolism [e.g. 43]. Each neuron of an afferent chain of neurons *encodes* a signal and transmits it to its target neuron or neurons where the signal is processed, initially by synaptic mechanisms.

As the data are propagated upward from neuron to neuron, relevant features are extracted and recombined by decoding and coding. Accordingly, the neuronal representation of graded information and its contextual meaning changes systematically from stage to stage.

Many variations of "neuronal coding", of the *how much,* are known. There is the simple rate-code, the temporal code, the cross-correlation code, the population-correlation code, the impulse-coincidence code, the position code and others [e.g. 78, 93, 109, 114, 124].[52] Notably the various codes all employ the all-or-nothing unitary pulses (spikes) separating time intervals. At least in primary sensory neurons the intervals designate (the inverse of) stimulus intensity or its rate of change, nothing but aspects of the *how much.*

The *Principle of Undifferentiated Encoding* [49, 120] states just this, and insofar it is disturbingly correct. However, as will be shown, the principle is incomplete and errs when claiming that the *what* and *where* is guesswork.

---

52 It is obvious that most of the above terms refer to processing rather than to coding.

Neuronal Identity and Targeting: Regarding the fixed or contextual part of the data, it should answer questions about the *what* and *where*. Such not-graded information can hardly be represented by pulse patterns. Rather, since a neuron is increasingly recognized to possess a distinct identity [48], I suggest that the contextual part for instance of the primary sensory message is based on this identity of the primary neuron.

The identity has become evident in several ways, notably also by specific molecular markers expressed on the cell surface. In development, several markers may be active sequentially [129]. Identity is also evidenced by the specific synaptic contact formed upstream with the next stage neuron.

To give an example, the targeting of outgrowing axons of olfactory receptor cells shows that the connectivity of a sensory neuron within the neuronal system is carefully guided. It clearly depends on the gene for the particular olfactory receptor expressed by a sensory neuron [48]. Different sensory specificity leads to connections at different target neurons. Both become reliable features of construction. They define the implicit context for the interpretation of neuronal spike signals.

Thus information about w*hat* and *where* is already contained in properties of the sensory neuron. It is due to matching connectivity to specificity in an autopoietic process. This process seems to realize part of the construction plan of our neuronal system.

Yet, the present wisdom of the field discusses data-codes of single fibres in terms of spikes, spike rates and analogue spike intervals alone. *"Spikes are the language of the brain"* [page 1 in 114]. Similarly, both J.H. Martin [page 336 – 338 in 93] and C. Koch [page 333 ff in 78] hold that neural code is spike pattern.

However, spike pattern or interval pattern is merely the rapidly variable, graded part of the transmitted neuronal data. Unmentioned in the above references, but equally important, is the con-

5. Neuronal Systems

textual or "fixed" part of the data.[53] This contextual part consists of discrete molecular information (e.g. Lipscomb's "glycocode" [89] and Hong's "combinatorial surface code" [68]), revealing the identity of the neuron in question and the connectivity resulting from it.

In this sense the 'undifferentiated encoding' of constructivism [49], claiming that the *what* and *where* is guesswork, is mistaken. It is the myth of the missing code.

Only if both parts of *hybrid coding* [54] are recognized as necessary items in their own right is it justified to speak of a *complete* neuronal code. This code, then, is built like any other datacode, it consists of a variable part and a fixed part. Intensity is encoded by spikes and contextual *what* and *where,* the meaning, by molecular markers and neuronal connectivity.[55]

MMC: Myth of the missing contextual code.

The fixed or contextual part of neuronal data should answer questions about the *what* and *where*. Such not-graded information can hardly be represented by pulse patterns. I suggest that the contextual part for instance of the primary sensory message is based on the identity of the primary neuron. Thus information about w*hat* and *where* is already contained in properties of the sensory neuron. It is due to matching connectivity to specificity in an autopoietic process. This process seems to realize part of the construction plan of our neuronal system.

---

53 J.H. Martin notes for primary sensory data that *sensory modality* is, of course, encoded by a labelled-line design [page 338 in 93]. Thus the identity of the neurons is taken into account in a broad way. Note, however, that identity based on *receptor expression* resolves more finely than identity based on sensory modality. In the case of the human olfactory system it resolves at least 340-fold finer than the modality "olfaction".
54 „Hybrid" refers to the complementary use of (a) variable analog data and (b) fixed (discrete) molecular symbols or topological features addressing the context of the variable part.
55 Note that the variable part is analog-variable rather than discrete--numerically variable.

At every stage of processing the pattern triggers a stereotypical reaction or a reaction modified by memory. In many cases the reaction is predictable. A simple message may read something like this: "Neuron #10345, connected as specified, reports constant intensity of value 101." The connection plan implies that #10345 is a sensory neuron connected to a warm receptor in a certain area of the skin. The neuron reports a skin temperature of 20° C at that location. Thus the *what* and *where* is known by implication from the wiring plan, providing the contextual part of the data.

Generally a clue to context and meaning may presumably be found from the total weighted input connections of a target neuron, including, where applicable, input from memory. An analysis leading up to meaning, not included in Shannon's 1948 theory of information, is an important goal in the ongoing investigation of neuronal codes.

5b. A hierarchy of neuronal mechanisms

Here the focus is on the attribution of models of molecular and neuronal mechanisms to system levels, loosely guided by the Churchland-Sejnowski system of levels in neuroscience [28]. 'Levels in neuroscience' is a subtopic of 'levels in science'. The levels may be ordered by organization, by processing or by something else, but often fall into an order by size.

In my view system levels are our pragmatic conveniences rather than given by nature. They serve for economy of thinking. By using a higher level, detail known but unnecessary on this level is hidden. Since the working memory of humans can handle only 10 items at once (or less, Miller's number [e.g. 5:259]), such economy is badly needed. [56]

Attention is paid to an arguable linkage of levels by bridging relations, a point not explicitly discussed by Churchland and Sejnowski. I hold that linkage or bridging occurs by prerogation, by *symbols* only. Each level symbolizes items of the level below

---

56 For a similar view, compare [134:3]

and ultimately of the zero-level, but groups them in a level-specific way (Figure 3.2 and 5.1).

Neuronal system level n minus 2 (n-2) harbours molecules. These are holons (see Figure 3.1 and [81]) constituted of (or described by) atoms, forces, fields and spaces. Modelled with their cyclic STDs, *single molecules* generate stochastic behaviour, while the SB of *ensembles* of molecules is more predictable though still fluctuating.

We may be especially interested in molecular ion channels, tunnel-like proteins spanning the surface membrane of neurons, serving for electro-diffusion of inorganic ions. A potassium channel molecule on level n-2 can assume one of several states, as specified in the STD of the model chosen. The STD is cyclic, supporting steady-state performance. Its ensemble SB is an irregular net-flow of K^+-ions, usually directed out of the cell.

The environmental energy source driving the cycle of states is a combination of potassium- and voltage-gradients across the membrane. Thus, as the net-flow has an effect on the K^+-concentrations, there is negative feedback from SB and EB to driving gradients. With millions of potassium-channels operating in parallel in one square-micrometer of membrane area, their ensemble-behaviour becomes quasi-deterministic. Therefore, with little error, SB and negative feedback via EB is quantitatively predictable.

It is important to realise that, in a conventional system, level n-2 harbours not just one type of channel or molecule, as sometimes depicted, but all channels and other molecules of the neuron, including those of synapses. Biochemists guess at 10^{14} or more protein molecules in a microliter of cell content (cytoplasm). Many of these molecules are interacting, the complexity is very large. This is precisely the reason why the next higher system level is needed.

System level n-1 harbours organelles and other multi-molecular cell components, such as input synapses, dendrites, axon hillock, axon, output synapses. On this level molecules are represented by symbols, abstracting from unnecessary molecular detail. As stated, the overall density of molecules in the organelles may be 10^{14} per microliter, or larger. The organelles show probabilistic to quasi-deterministic behaviour, depending on details of their model.

To give an example, the molecular level n-2 specifies ion channels. In the node-membrane of an axon these amount to thousands of Na^+-channels and thousands of K^+-channels, suggesting quasi-deterministic behaviour. On level n-1 the ensemble of Na^+-channels may be represented symbolically by a conductance g_{Na} and the ensemble of K^+-channels by a conductance g_K. The conductances are part of the organelle 'axonal surface membrane'. The gating of Na^+- and K^+-channels is voltage dependent and interacts by means of the membrane voltage common to all channels. The resulting time-course of membrane voltage is the action potential (the nerve pulse), modelled with *deterministic* differential equations [65] as interaction of g_{Na} and g_K in the axon membrane, an organelle of level n-1. Yet, unexpected from the deterministic equations, recordings from an axon at high sensitivity reveal pronounced fluctuations of membrane voltage.

Fluctuations on the level of organelles were especially noted for cortical synapses, were transmitter release is highly probabilistic, as detailed below (see Section 5d, 'Randomness').

On system level n we focus on the various neurons, i.e. holons constituted of organelles. On level n the organelles are represented by symbols, hiding unnecessary detail. In a neuron may be in the order of 1000 synapse-organelles, most of them input synapses. Yet, there are about 10^{11} neurons in a human, many of them interacting with each other via synapses [e.g. 79:16]. These 10^{11} neurons are all assigned to level n, again the complexity is large. However, we usually single out only one or a handful of neurons when working with level n. Each contains its organelles and its functional cell parts as symbols.

The STD of neurons specifies a steady state cycle, but with transitions triggered by the input. For instance, a neuron may be in a resting state but will transit to states of excitation or inhibition, depending on synaptic input driven by other neurons. The neuronal model specifies which output-SB (pulsed secretion of transmitter molecules) results from a given synaptic input (pattern of transmitter acting on post-synaptic receptors). When input ceases, the resting state is re-established. Note that the output spike pattern especially of cortical neurons can be stochastic, as detailed below (Section 5d).

On system level n+1 we collect network-mechanisms, i.e. mechanisms involving several neurons. The neurons are represented by symbols, abstracting from here unnecessary cellular detail. The number of possible networks is large, since overlapping functional coalitions of neurons may be formed in the same neuronal population.

To give an example, for a network-mechanism - in a simple case known as lateral inhibition - an array of neurons shows an enhancement of cell to cell output contrast, if each neuron inhibits its neighbours to a degree proportional to its own excitation. Here the neuron-symbols carry but few properties, like gain and spread of excitation and inhibition. The SB is the pattern of secreted transmitter across the array of neurons, indicating the exaggeration of contrast. The network-model predicts to which degree this exaggeration will occur.

A word of caution: Despite the impressive success of detailed biophysical research on networks of neurons, the goal set has not been reached. Reviews show that it is almost impossible to understand the multiple-feedback situation of complex networks of neurons. The six-layered iso-cortex or the basal ganglia provide striking examples [e.g. 118]. This calls for computer simulations [92]. They help to integrate our bits of detailed knowledge into a hypothetical functional whole. However, the result must be broken down again into palatable chunks which may be comprehended by humans. The limitation of our cognitive ability is hard to overcome. Presently we fail to understand even a simple thought in terms of neuronal mechanisms.

Specifications and limitations: The specifications of mechanisms are also their limitations, often resulting from forced compromises in design. For instance, speed of cortical processing, clearly an advantage for survival, requires fast conduction along short connections. This means high connectivity, i.e. high packing density [e.g. 116:350ff]. These requirements are contradictory since at high packing density the smaller axon diameters result in slower conduction. Further, small neurons have higher processing speed but are noisier and have less metabolic and respiratory reserve. Also, high packing density demands higher energy throughput and more efficient heat removal by perfusion with blood. This, however, requires volume, decreasing packing density.

The human brain is the largest among those mammals having a 'thick' iso-cortex (the human cortical width is 3-5 mm). Further, it has the largest number of cortical neurons, more than 11000 million. Its frontal-lobe volume is 280 cm^3, more than threefold larger than that of other primates [e.g. 116:343ff]. Otherwise there is no molecule, neuron type or principle of function known in the human cortex which would be typically human and would explain a superior position among mammals or primates. Only cortical packing density is slightly improved in humans. Cortical volume and packing density seem to be important limiting parameters.

Generally we suspect that any performance of our neuronal networks has its limitations, PLC of working memory (Section 5c) is only one among many.

We may continue to build our stack of system levels *bottom-up*. Still following Churchland and Sejnowski [28], possible stations are neuronal maps, sub-systems (neuronal modules), the complete central neuronal system (brain and spinal chord). From there it is tempting to continue with 'coalitions of communicating brains'. This would allow to discuss shared concepts, content of the sciences, art and culture.

So far the conventional levels of organisation and processing are also levels of size. Since they are all conceived as 'levels in neuroscience', items like 'geology' or 'planetary system' do not occur, except as content of communication of brains. They have to be dealt with by a hierarchical system of their own, like 'levels in geology', another topic of 'levels in science'.

But where is the *mind* in the neuroscience scheme, does it occupy a level by itself? This will be discussed below (Section 6k). Note, however, that the mind, Descartes' *res cogitans*, has no physical extension, is immaterial. It cannot be accommodated if size is used to order the levels.

5. Neuronal Systems

5c. Universal levels of neuronal mechanisms

Above, neuronal mechanisms of increasing complexity were organized into a hierarchy of system levels. Holons (mechanisms) on one system level were represented on the level above with symbols (wholes) and these were grouped into higher-order mechanisms on this level. With universal system levels, USL, the vertical coupling of levels is neither causal nor constitutive, but by identity of the basal items (Section 3f).

Such *universal* levels, being all-comprising, necessarily contain the same basal physical items and events. The difference is in their ordering, which is realized by assignment of symbols to level-specific groups. Thus, while basal events occur synchronic on each level, they are noted as changes of different higher-order hierarchical structures. The idiom describing these structures and their relations necessarily differs from the idiom describing the basal events. The connection may be difficult to recognize. Here is an example:

At time t a neuron fires a spike. This would be an event at level n. It is due to a series of organelle events involving current loops through membrane areas of the neuron, events at level n-1. These are due to molecular changes at ion channels and ion pumps, events at level n-2. They, in turn, are due to sub-molecular events like allosteric reconfiguration, movement of gating charges etc., events at level n-3. The sequence of levels may be followed downward until level-0, the basal level is reached. However, a description of the spike in terms of basal events will be very difficult to comprehend.

PLC and decreasing explanatory value in multi-level reduction: True, given unlimited cognitive ability, events at level-0 would explain the spike at level n in a subjectively convincing way. But the seemingly strange nature of basal events, their huge number and the idiom associated with them makes it highly impractical to describe the spike in terms of level-0. Thus, pragmatically a series of translation-levels is inserted upward of level-0. Clearly, they are needed for epistemic reasons, since they reduce the number of items (symbols) to be considered at once. After all, as stated, our working memory can handle few items simultaneously (Miller's number [5:259, 96]).

Therefore, single-level reduction (relating a level to its nearest neighbour, below) may be subjectively convincing but multi-level reduction is not: By skipping intermediate explanations it necessitates the handling of too many items.

> PLC: Principle of limited cognition.
>
> The brain's abilities are large but necessarily limited. For instance, the working memory of humans can handle only 10 or less items at once. Therefore, we create upper system levels, where fewer items (symbols) need to be considered as multiple components are represented by one symbol. Looking downward, multi-level reductions tend to have diminishing subjective explanatory value, as more intermediate explanations are skipped. Then relations may be objectively proven by science but not understood subjectively.

To obtain *universal* system levels, we project the basic level vertically upward, showing that each level contains the basal events. They are simply grouped differently and presented in different context and idiom as we move from level to level. The ordering of mechanisms in level-specific units situated on universal levels is illustrated in Figure 5.1. Note that level n allows us to deal with three objects or events (star-shaped neurons) rather than with millions and trillions of basal objects or events (black dots constituting the neurons). Three but not trillions can be processed in working memory. This restriction indeed supports the claim of post-reductionists that our world view is co-determined by the limited cognitive abilities of our neuronal machinery [69:151].

Science, of course, is beyond a comprehension limited by PLC of working memory, it relies on objective proofs. To obtain them may be restricted by the brain physiology of our best thinkers. To accept them is a matter of the proofs themselves, our comprehension being of secondary importance.

5. Neuronal Systems

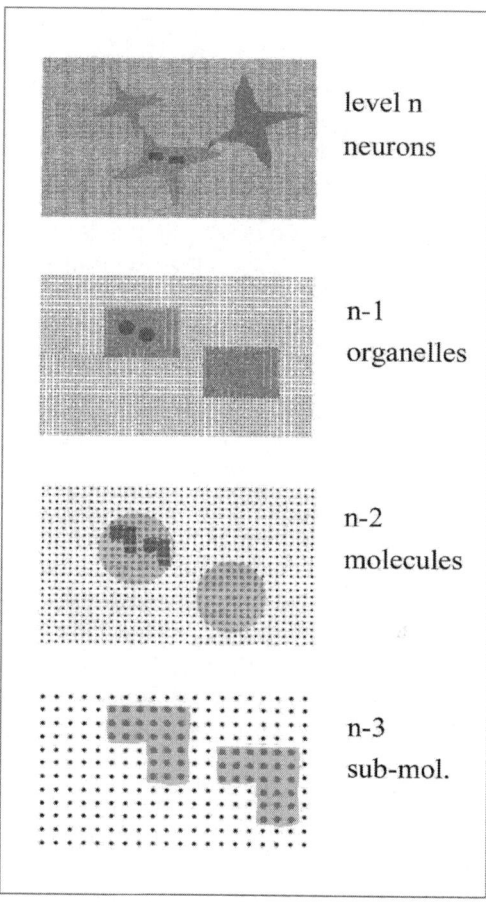

Figure 5.1. Clippings of some universal system levels in neuroscience. Meaning of symbols: **Stars:** events in neurons, spiking; **squares:** events in organelles; **circles:** molecular events; **angles:** sub-molecular events; **dots:** basal physical events. Note the change in scale as we move upward, zooming out. Complete coupling of levels is implied since synchronic events at two levels root in the same basic events which occur on each level. *Proper grounding* is a prerequisite.

5d. Randomness

As the grouping specifies mechanisms of increasing complexity and size, these become macroscopic and are often modelled as deterministic, thematizing ensemble behaviour (sum of single-unit SB) but not the fluctuations associated with them. Therefore it is to be noted that fluctuations remain associated even with large populations. Suppose the single-unit SB of a population of N parallel units is added. Then with increasing N the fluctuations of the ensemble increase (even though less so than the sum of SB). Thus the probabilistic fluctuations do not disappear when we change to a deterministic model of many parallel units.

Yet, as discussed below (Section 6e "Causality-gaps"), the mere presence of random fluctuations does not mean that causality is punctured. For even events of a Markov sequence may be modelled as a gap-less causal chain (Section 2e). The random sequence is constrained by the rules of the STD.

Passing through a network of neurons and neuronal modules we may envisage a sequence of signals fluctuating more or less at each stage. This is not a patchwork of alternating deterministic and probabilistic natural processes (as sometimes claimed [e.g. 46, 47:11]), but a sequence of probabilistic models, specifying randomness adjusted to different degrees from stage to stage.

The virtue of randomness: In the biochemistry of molecules in small number, fluctuations may even exceed the mean SB. They can increase sensitivity, allowing stochastic focussing, with noise facilitating signal detection in nonlinear systems [108]. In neurobiology, transmitter release at cortical synapses typically operates with few synaptic vesicles and a low probability of vesicle discharge, it is highly probabilistic [45, 78:90, 82]. Presumably the design of a random performance is of advantage, as on other locations very reliable synapses are also in evidence. Famously, synaptic strength is use-dependent [59], it is subject to long-term potentiation [78:317]. But not only synaptic signal intensity, also the reliability of vesicle release seems to be varied as a *feature* in the process of learning. It appears that a point on the scale from probabilistic to to deterministic is chosen by design. Transmitter release of baseline probability 0.1, say,

5. Neuronal Systems

would be increased with use to a probability near unity, "... lack of reliability at low intensity is required to give a synapse its large dynamic range" [78:327].

Randomness is continued in units of higher organisation [45]. It is remarkable that cortical neurons can show a completely random spike pattern as their typical output. In principle the degree of randomness can be tuned by putting more or fewer stochastic elements in parallel, such as ion channels, synaptic vesicles, synapses or neurons. In each case randomness can be varied, it is a 'feature'. In the case of cortical neurons randomness (or removal of redundancy) was suggested to constitute a new dimension for coding with limited bandwidth: reliability increases with signal strength [78:373]. Again, low signal strength is associated with pronounced randomness and high strength with predictability. Nevertheless, population averaging across many parallel neurons can improve overall reliability of the neuronal mechanism. That may be one reason why the human nervous system needs such a large number (about 86 billion) of neurons.

> Thus randomness is a feature of coding neuronal signals.

Has it other advantages, is it related to unpredicted behaviour, perhaps to fantasy, to creativity? It seems possible. A neuronal selector may randomly choose among a number of more or less well established possibilities. Fantasy amounts to thinking off the beaten tract. Not the population mean is processed, but something offset from it, appearing as a new idea. The neuronal basis of creativity will be a fascinating topic of future research.

6. Mind

The 1st-person perspective of mind is that of a conscious Self-agent experiencing the content which is displayed by abstracting from neuronal mechanisms. The format of this experience is that of multimodal scenes on an 'inner theatre stage' or workspace. The scenes serve for communication to neuronal modules and for their coordination (B.J. Baars). Since experience and thought are abstract, they are non-interacting on the mind-level. Nor is such interaction of mind → neuron needed, complete causal chains are already in the neuronal modules.

In the 3rd-person perspective the mind is a symbol for a bundle of experienced mental processes (including the psychological predicates) generated by neurons. For reduction, each experience of the predicate-processes separately is to be reduced to a neuronal process, taking it as a 3rd-person explanandum (Red B). The mental predicate-process and its neuronal base should have the same causal role. Further, there should be bottom-up manipulability.

Along these lines the mind-body problem and further aspects of mind are discussed.

Recall that the *world* is our physical environment, (in the idiom of level-1) a system of physical objects of matter, made of components, again physical objects of matter, energy, forces, fields, spaces. These material objects are ordered in space and time. They can meet each other in space and time and interact. They are not 'about' something.

Living beings are more than physical objects, they have design and agency. They also have a body, which is (in case of humans) 'owned' by a conscious Self. The Self and its body is part of the world, but the Self perceives itself as separate from the world, its primary perspective is one of agent-world polarity. Further, it is not aware of its body's neuronal system. This is transparent to the Self, though part of the body and of the world.

Yet, when viewed objectively in the 3rd-person perspective, the *mind* is a bundle of experienced 'mental processes' which depend on and may be a result of specific neuronal (physical) interactions in our brain. These arguably generate agency and the conscious experience of thought content by means of sub-agents working sub-consciously. Thoughts are immaterial, are abstracted from space and time. Therefore, it will be argued, they cannot physically interact. Further, like language, they are 'about' something [21].

When considering the mind, we enter territory which is very familiar to every human being as her/his interior view, yet difficult to evaluate objectively. The following chapters will discuss some highlights of our mental dimension, but, of course, cannot do full justice to the topic.

6a. Agent-world polarity

Sensory realism: A sensory organ of our body responds best to its *specific stimulus*: hearing responds to stimuli of the acoustic spectrum, vision to a narrow band of the electromagnetic spectrum, the sense of smell to some kinds of, but by no means all kinds of airborne molecules.[57] Thus the more or less continuous properties of the environment are monitored within certain windows only. Not to get all information is the point, but merely to get the information relevant for survival. Mercifully, we do not miss what we cannot sense, we are not aware of the wide gaps

---

57 For instance, we cannot smell oxygen or water.

6. Mind

in our perception. Yet, to fill the blind spots of our primary models, the scientific study of world and mind aims to generate an objective, rational and complete world-model based on reproducible empirical evidence.

Concepts about the world around us require a counterpart, a concept about ourself as an agent acting in the world. Thus the world model arising from sensory experience is paralleled by a primary Self-model which is indispensable even though subjective, of uncertain scientific reliability. Again we do not miss what we cannot sense. In particular, in the interior view we are not aware of our neuronal system because we have no sensory organ to note our own neuronal activity.

> STN: Subjective transparency of the neuronal.
>
> The neuronal is transparent in the sense of 'present but unnoticed', like glass to vision. The activity of our neuronal system is subjectively transparent because we have no sensory organ for such activity [e.g. 87:240, 95:188,289, 97, 104:267, 105:95]. 'Through' the unnoticed neuronal system we sense, by neuronal function, features of world and body. And we do not miss what we cannot sense. This helps to establish the notion of a mind and Self existing independently from the neuronal-physical world.

Primary models: How do we group sensory phenomena into objects? The rule is: What correlates, belongs together. This principle, doubtful or downright wrong in science, is of great practical value as a fast-and-dirty rule for the interpretation of sensory data. For instance, in vision those image parts which move together, are treated as belonging together, as belonging to a particular whole. For such strong correlation of movement likely indicates a world-object distinct from other objects, whose parts may also move together, but differently. Thus the parts-whole relation is a principle for the interpretation of sensory data.

Empirical dualism: The infant, exploring correlations while learning the identification of objects, will notice that there are objects whose parts also move together but their movement can be wilfully initiated. Further, these objects 'feel', they are the origin of special sensations. After much experimenting the child assigns such objects to the agent which claims to be the cause of the movement, the conscious Self. Such objects are parts of the child's own body, distinct from the world. The Self experiences itself as the body and the world is outside from it. The dualism of world and Self-agent is born (sensorimotor stage of Jean Piaget).

Further, those perceived movements of world-objects not claimed by the Self are assigned to other agents, fictive agents which are assumed to be out there. The infantile world becomes animated, magical (preoperational stage [110]). While the animation of world-objects is corrected in later stages of child development, the animation of body-parts by the Self is retained.

Confidence in our primary world- and Self-model, field-tested and indispensable yet doubtful as it is, bears the philosophical label "folk-psychology" [26, 27] or "naive realism" or "primary realism", meaning that such confidence is based on uncritical trust in our primary experience with our world and with ourselves. It is, of course, the task of science and philosophy to probe and evaluate such models.

6b. *Bottom-up.* Do biophysical mechanisms give rise to mind?

When probing primary models, rather general questions arise. Can neurons actually generate abstract thoughts, can phenomena of the brain 'give rise to the mind'? Objectively, the mind's existence and detailed function seems to be necessitated by the proper action of neuronal mechanisms in the physical world. Also, subjectively, daily experience with our body, with sleep, wakefulness, drugs, hormonal states, cerebral blood circulation etc. leaves no doubt about this dependence. In terms of conventional systems theory such supervenience of mind over body is accounted for by placing mind-phenomena on a system level above the level of brain-phenomena.

Mind-phenomena, I suggest, are conscious *experiences* of an interior view, showing how our Self is positioned in the world in past, presence and future. Brain-phenomena, in turn, are events due to cerebral neuronal networks functioning *biophysically* as their building plan stipulates, underpinning the mental experience. How, then, is mind related to biophysical mechanism, how are such system levels coupled?

Again my suggestion is, contrary to conventional systems theory, that a special vertical coupling of levels, causal or constitutive, is not needed. *Universal* levels containing synchronic events rooted in basal events are sufficient (Section 3f).

When we experience a neuronal event in the presence, or when we experience a past event from memory, we refer with a symbol to the *content* of the neuronal message which reaches the consciousness-modules of the brain. This symbol in the mind-level is a conceptual stenogram for the content of the neuronal circuits employed.

The mind-level m, then, contains an assembly of symbols referring to content of messages on the neuronal level m-1, below. The content on level m-1 is neuronally registered on level m-1 and mentally experienced *bottom-up* with symbols on level m. The *symbols appear to interact*, but this is only a projection of the real interaction taking place in the causal chains of the neuronal biophysical world. Yet the apparent interactions of symbols seem to follow their own syntactic rules, the idiom being that of the mind-level.

The mind-idiom is talk *about* content of brain messages. Thus, clearly the mind depends on the brain like a symbol depends on its denotate. This is the *bottom-up* dependence. However, will the mind-symbol also have a *top-down* effect on its denotate? Note that generally a symbol depends on its denotate but the denotate does not depend on the symbol. There is an asymmetry of existence.

Yet, in everybody's subjective experience mind activities like thinking have causal relevance, they have results in the physical world, like writing on paper. Does the mind affect the brain, the hand? Or is it one brain module which affects another brain module at level m-1 and the process is experienced with symbols at level m, giving rise to the illusion of a mental syntactic manipulation of symbols? This would mean that we experience an illusionary *top-top* process while actually a *bottom-bottom* process is taking place.

6c. *Bottom-up* and *top-down*. Thoughts control neurons?

The above problem may be focussed by asking whether our thoughts actually control body movements (this would be via neurons), as suggested by our Self-experience? What is a thought? It is a mental operation with concepts. It is a try-out of action with reduced risk (S. Freud). It is acting in virtual space [90:175, 111]. Further, more technically, it is the experience, the awareness of content.

Thought is experience of the content the thought is 'about'.

The deeper question is, what is 'content'? In a way, content is the ghost in the machine. Something immaterial, not located in space and not located in time. Removed from space and time it is an abstractum. Being immaterial, it cannot interact physically, it lacks causal power. Thought and content, then, cannot control neurons. Something immaterial cannot interact.

ANA: Abstracta non agunt in concreto.

Abstracts are not locatable in physical space, they cannot interact physically, are without causal power. To attribute causal power to an abstractum means to make a category-mistake, a reification, it means to fall for the fallacy of mistaken concreteness [139:51]. Thus abstract thoughts (immaterial thoughts) cannot cause.

6. Mind

Let us probe how content is handled generally in communication. There always seems to be

(a) a gap-less causal chain of physical events, producing a pattern,
(b) interpretation of the pattern as known symbols by the mind-brain and
(c) association of the symbols with their implemented or memorised or synthesised denotate, the meaning, which is content.

Take a computer. It is advertised as a machine manipulating binary numbers. But actually a computer's processing unit is a hopefully deterministic mechanism which does not manipulate abstract numbers but concrete electrical charges. Computing is physically a gap-less causal chain involving patterns of charges (a). To the inaugurated programmer such patterns are symbols of symbols (b). It is the programmer's brain which translates the secondary symbols into numbers and the numbers into meaning. Presumably this happens by association of the symbols with their memorized denotate in the programmer's brain (c).

Or take the so-called transmission of information. Information is an abstract, it is immaterial. Can it be transmitted to remote locations? No! [e.g. 22]. Rather, it is some physical entity which is transmitted in a sequence forming a pattern. And it is the receiver, our mind-brain, which interprets the pattern as symbols and associates the symbols received with their memorized denotate, the message.

Similarly, a neuronal thinking mechanism generates a gap-less causal chain of biophysical events. With the appropriate neuronal code these events are (by other neuronal processes) interpreted as physical symbols which have a meaning, their content. The content is already in memory due to other, previous biophysical events.

Now about *top-down* linkage: Those biophysical events have causal power but the abstract content does not. Immaterial abstracta cannot interact. Thus abstract thoughts cannot influence neurons. There is no way down.

Of course, it is possible to define 'thought' not as the experience of abstract content but as the biophysical mechanism which allows this experience:

> Thought is the neuronal mechanism which allows the experience of abstract content.

Then thought can influence or control neuronal activity because it *is* neuronal activity. But why do we shake our heads? Why is this simple solution so counter-intuitive? We prefer thought to be immaterial because we are blind to our own neuronal activity and this blindness has formed our intuition. We perceive only what sensors reveal to us and we lack sensors for the presence and the interaction of our neurons. These, therefore, are subjectively transparent.[58] And, indeed, we don't need such perception. It is sufficient that the neuronal mechanism works, what we need to perceive or experience is its content only.

6d. Gap-less causal chains

Above it was claimed that neuronal causation occurs in a gap-less sequence in which thoughts and other immaterial abstracta cannot participate. This is also true in general, physical causation occurs in a gap-less sequence in which immaterial abstracta as such cannot participate. What, then, is so convincing about statements like

> "An error in the software caused the rocket to explode",

in which 'software' is an abstract and 'explode' an event of the world? To probe this difficulty further, let us take computer programming as an example.

The computer hardware is a machine made of mechanisms which may be modelled deterministically. However, as noted before, computers, the "number crunchers", do not crunch numbers. Rather, it is patterns of physical charges which are guided into causal chains by the computer hardware. The patterns stand

---

58 Subjective transparency of the neuronal activity (STN). Present but unnoticed, like glass to vision [e.g. 95:188, 97].

for commands or for data. In common wording, the causal chains are implemented by *coding* with low-level software (binary machine language, compilers). Yet, there is no causal path from software to computer hardware. Rather, causation is from a tape ('containing' the software) to the hardware. The pattern on the tape, in turn, is not caused by software but by current through a magnetic writing head. This current may follow a pattern produced by a keyboard. Its action, in turn, is dictated by a human hand. From here the causal chain may be followed backward into the realm of neuronal interaction, again a gap-less sequence of physical events.

Notably, a causal path from abstract, immaterial software to concrete hardware is not found. Rather, the path leads from material patterns (on a tape, perhaps) to hardware. The patterns represent the software and can be scanned physically, attaining causal relevance.

For *decoding* of the computational result, the interpretation of patterns of physical potentials as numbers and of numbers as other symbols (using code tables) is achieved by neuronal biophysics. Then the symbols are *associated* with conventional meaning, which was at a previous time consciously experienced and stored in memory, again involving neurons in a gap-less chain of causation.

It remains to see how the association of a symbol with a denotate, the message, comes about. The association and the message was learned at a previous time. *Learning* means that (1) a sequence of low-level physical patterns like letters or syllables is first parsed into words which are lexically associated with low-level meaning, i.e. memorized verbal concepts. (2) Then their combination to more complex concepts (like the meaning of sentences) must be 'understood'. Such understanding or comprehension occurs by reconstruction with syntactic rules or by reconstruction with other rules of synthesis.

Message reconstruction along rules is a necessity because its alternative, the lexical look-up table, is of limited use if it becomes too long. Once the message has been learned, it can be retrieved from memory by association. In common wording association, learning and reconstruction are 'mental processes', or appear to be such processes. They are without causal power but have a neuronal basis and this alone has causal power. In short all processes mentioned are neuronal, ordered in gap-less causal chains (see Figure 6.1, below).

In this vein, the 'error in the software' caused nothing but had a material basis which, by being scanned, attained causal relevance to other events which, alas, 'caused the rocket to explode'.

6e. Causality-gaps

According to physical experience, non-physical influences cannot interfere with the causal chains of the physical world.

This principle was recently challenged by postulating so-called 'causality-gaps'. The partial randomness of neuronal processes was suggested to puncture the causal closure of the physical world [46:378, 47:17]. Through the gaps immaterial thoughts were proposed to influence material neurons. However:

1. According to Section 2b, the mesoscopic physical level with its probabilistic descriptions is rooted in the microscopic level (more realistic but more difficult to handle) with deterministic descriptions. Where fluctuations and randomness on level-2 are due to $1 \rightarrow 2$ simplifications, they do not exist on level-1 and, therefore, cannot provide for "gaps".

2. As outlined above (Section 2e, *Causal models)*, probabilistic models are causal, too. The direct cause for a stochastic event is that a random variable exceeds a threshold (at an on level-2 unpredictable time). Stochastics alone is not evidence for causality-gaps, as the cause is found in the event of exceeding a threshold.

3. Further, such presumed puncturing of causality by immaterial thoughts has a plausible alternative: that thoughts have a neuronal basis through which causal interaction occurs [e.g. 74:92, 88]. This conventional alternative to causality-gaps is to be considered.

> CCP: So-called "causal closure of the physical world".
>
> The expectation, based on experience, is that every physical change is caused by physical phenomena. Every action on matter is interaction with other matter (a consequence of Newton's third law). Then non-physical influences cannot interfere with the causal chains of the physical world. The reason is that non-physical entities are necessarily immaterial, therefore do not interact with material objects, cannot encounter them in space and time. In consequence, an immaterial mental phenomenon (like a thought) cannot 'cause' physical changes at neurons.

It appears that Whitehead's "fallacy of mistaken concreteness" [139:51] continues to trap authors. To give another example, it was argued that there obviously is a causal relevance of immaterial reasons. A reason, which I consciously deliberate, then guides or causes my action, or contributes to it. This was taken to speak against a causal closure of the physical world [103:257]. Yet reasons, being immaterial, cannot interfere with a physical causal chain. Rather, it must be the neuronal basis of the reasons, their reduction base, which has causal relevance and causal power. And this basis, of course, is part of a causally closed physical world.

6f. Relation of world and mind

Above level-0, the physical world is "res extensa", matter and its interaction. Mind is without location and extension, is immaterial experience, view, not physical events but *about* physical events.

The human mind may well be a result of neuronal interactions in our brain, which generate first of all the experience of an interior view. This view shows our conscious Self, an agent experiencing the mind phenomena, positioned in (the remainder of) the world, in past, presence and future. The mind and Self, apparently existing separate from the world, are abstractions from the world.

These abstractions are of great conceptual consequence, though illusionary if the mind (and Self) is understood to exist independently and act independently [e.g. 137, 138]. Rather, observations in the third-person perspective shows the mind's existence to depend on the action of neurons in the physical world. These, however, remain subjectively transparent because, as stated, we have no sensory organ to note our own neuronal activity.

Can mind have mechanisms? Suppose somebody has thoughts and writes them down on paper. There is a causal chain of neuronal, biophysical and physical events which without gaps constitutes this activity. In the framework of a conventional hierarchical system we may call this activity *bottom-bottom*. Further, the thinker has an experience, he becomes conscious of the content and its change, of a flow of content. The flow is due to a sequence of discrete *bottom-up* events.[59] However, since the bottom dimension is mentally invisible, the sequence will appear to the thinker as a *top-top* flow of content.

The mental contents seem to interact, their changes seem to follow rules. This is entirely due to the *bottom-bottom* mechanisms but, since these are transparent, it appears as immaterial mental activity following its own rules, seemingly based on mental mechanisms.

Now suppose the neuronal thinking process solves a problem. This occurs *bottom-bottom*. The solution will appear *bottom-top* as experienced content. Yet, since the bottom dimension is transparent, it will look like a *top-top* solution in a *top-top* flow of content, a product of immaterial thought. Similarly, when the pen is wilfully moved, a *bottom-bottom* or neuron-hand-pen event, this will look like a *top-pen* event since the bottom dimension is transparent.

Thus the neuronal material base is experienced as immaterial,

---

59 These bottom-up events have a pronounced time requirement of 0,3 to 1 sec [115:133].

6. Mind

the *bottom-top* origin of mind-phenomena remains hidden. In consequence, mechanisms of the mind alone appear to exist but are illusionary. What they seem to achieve is due to neuronal mechanisms.

A presumed *top-down causation* of mental events may be analysed as follows: There is a *bottom-bottom* causal chain $n6 \to n8 \to n10$ and the especially time-consuming *bottom-top* event $n6 \to m6$. Now any observed variation in $m6$ will be followed by a variation in $m10$ and $n10$. It will seem that $m6$ is causing $n10$ (and subsequent body movements). However, $m6$ cannot be changed independently, only by first changing $n6$. Further, inhibition of $n8$ will interrupt $m6 \to m10$. Thus the causal power of mental events is only apparent.

The relationships are summarized in Figure 6.1. The diagram describes a conventional hierarchical system with a *bottom-bottom* chain of interactions "along the roots", and a *bottom-up* linkage of levels. The upward linkage stands for conscious experience of content. Notably, for the reasons given on the right, there is no causal linkage back from mind to neuron: the immaterial (experience, content) lacks causal power but is *'about'*. Like a language, also immaterial, mental phenomena are *'about'* something [21:124]. This 'something' is already processed in the roots, it does not require *top-bottom* linkage.

Turning to a system of *universal* levels, here identical basic physical events are found on each level. Only their grouping (and assignment of symbols to groups), context and idiom changes from level to level (Figure 5.1). The upper sequence $m6, m8, m10$ is a sketch of the lower sequence $n6 \to n8 \to n10$ with symbols representing content contained in the lower sequence and obtained from it by re-grouping. Steps like $m6 \to m8$ are symbolic events based on $n6 \to n8$. Reduction to biophysics is possible along Red B (Section 4b), viewing the experience of the upper sequence as an explanandum in the objective 3rd-person perspective.

> Generally, mind-phenomena are products of brain-phenomena highlighted by the illusion that brain-phenomena do not exist.

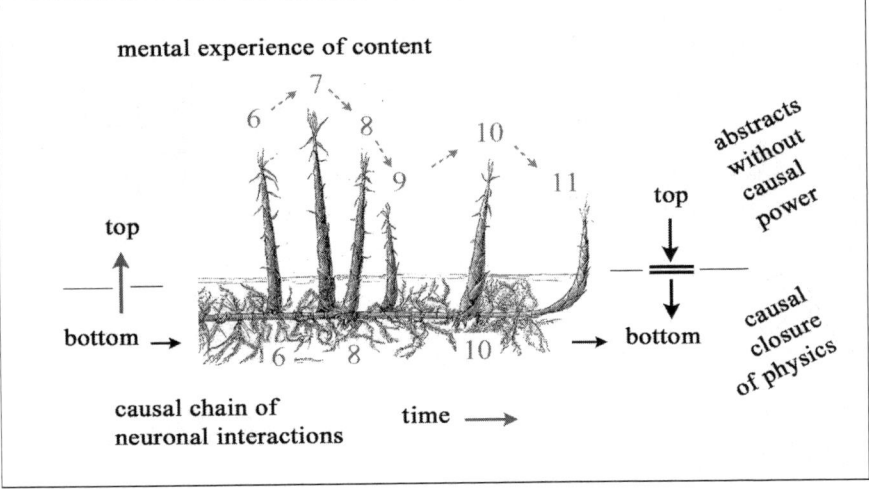

Figure 6.1. Mind-neuron relation as a conventional hierarchical system where mind supervenes neurons. The horizontal sprouting of a bamboo plant illustrates the *bottom-bottom* flow of neuronal interactions along a gap-less causal chain. These are the neuronal 'roots'. Discrete *bottom-top* links (neurons generate thoughts?) lead to mental experiences of content. They are consciously perceived as a *top-top* flow of content (dashed arrows), as the *bottom*-dimension is transparent. *Top-bottom* linkage (thoughts control neurons?) is impossible for several reasons (two are stated on the right, a third may be the subjective transparency of the neuronal system, STN). Nor is such *top-bottom* linkage necessary, since the complete causal chain is already contained in the neuronal circuits.

6g. Extension to psychology, sociology etc.

The representation of holon-mechanisms by symbolic wholes and the explanatory reduction of symbolic wholes and their SB to holon-mechanisms is a flexible framework which may be extended beyond the Self to shared fields of knowledge like psychology and sociology. Psychological predicates or 'mental processes' (e.g. thinking, feeling, remembering, judging, deciding, acting, behaving [14]) are research topics of psychology. Here, too, each predicate is a symbol which stands for a mental experience (1st-person) based on a bundle of complex and specific neuronal events (3rd-person perspective). The same holds for concepts like parenthood, bravery, dissatisfaction, happiness, qualia, agency, dignity.

About reducing dignity: The reduction of agency was discussed in Section 4g, let us take dignity as another example. Can dignity be explained by neuronal mechanisms? The case is of special significance since 'dignity', like agency, is often quoted as a property of humans which is clearly 'more than physics'.

Dignity is maintained when we act, speak or think such that certain restrictions are met. For this we are rewarded by feeling valuable in a particular personal way. The restrictions may have been communicated from the human environment by the example of peers. To reduce 'dignity' means to reduce 'acting under restrictions'. This needs, say, a list of goals and do-nots, further a decision mechanism and a motor-executive device. There is no doubt that such functions are available in our neuronal system. Therefore the first-person predicate-phenomenon 'acting dignified' may likely be reduced to third-person neuronal mechanisms. This would be a one-level reduction with clear explanatory value. Further reductions lead down to physical base, but comprehension may be lost on the way.

Sociology: Psychological predicates were taken to be symbols for the mental experience (1st-person) of processes and concepts based on a bundle of complex but specific neuronal mechanisms (3rd-person perspective); this relation may less directly also hold for concepts of sociology like solidarity, worker-power, population, capital, morale, culture and so on. Note that these are abstracta, where in space is 'culture' or 'solidarity'? They are symbols of symbols of symbols... arguably based on neuronal events. The peculiarity is that the conceptual meaning of these symbols is shared by a community of people. The concepts have no independent existence but depend on thinking and memory of the members of this community.

In the end all these concepts, in fact all concepts, may be reducible to combinations of neuronal events which, in turn, are physical events or based on them. However, I hasten to add, there is really no need nor use to undertake such a tedious multi-level reduction. It means to omit intermediate explanations while these are needed for comprehension. After all, system levels are inserted for epistemic reasons, parsing phenomena into a few handy chunks which can be processed together (PLC).

Thus the solution of problems of sociology will hardly be furthered by the knowledge that this science is - very indirectly, through the neuronal base of mental concepts - based on physics. Yet, there is no principal difficulty, reductive physicalism remains a hypothesis not falsified. And it still carries the promise of a unity of science.

6h. Mind immaterial

Here is a summary of my analysis. The mental is part of the first-person perspective, 'mind' refers to the *experiences* of the first-person, of the Self-agent. They are conscious experiences (multimodally communicated on the inner theatre stage) of control and causal power, further of so-called 'mental processes': thinking, feeling, remembering, judging, deciding, behaving (the 'psychological predicates' [14]). All these mental processes have a neuronal base but are experienced by the Self as activities of the Self.

Thus the human mind is a bundle of *experienced mental processes* (based on neuronal processes). Where in space is mind, is experience, is, for example, 'judging'? Nowhere, experience is an abstract, it is 'about' content. Having no spatial extension and location, the mind may be viewed as immaterial. Being immaterial, it is without causal power. Yet, the Self experiences causal power. Blind to neuronal activity (STN) it attributes to itself what is really caused by its neuronal base (Figure 6.1).

To mention another view, Jaegwon Kim rejects that mind is immaterial [74:91] because "it seems beyond doubt that mentality is part of the causal structure of the world". For Kim the term 'mind' refers to material phenomena, to which functions may be attributed. This, his causally active mind, will be the neuronal base of what others take to be the immaterial mind.

6i. Mistaken concreteness

Concreta and abstracta belong to different ontological categories, they must not be confused. Concrete objects are material physical objects which can be located in space and time. They can encounter each other in space and time and interact causally. Abstracta are immaterial, they are not locatable in space, they cannot interact, are without causal power. To attribute causal power to an abstractum means to commit a category-mistake, a reification, the *fallacy of mistaken concreteness* [139:51].

This mistake is very common. An example was mentioned above: *"An error in the software* (abstractum) *caused the rocket* (concretum) *to explode."* A more philosophical example is part of the Bieri-trilemma [18:9], which, in my words, runs like this:

There are three statements, touching on the mind-body relation, which seem plausible. Yet, they are contradictory and one of them must be denied to remove the contradiction. Here they are:

>   MdB = mind different from body
>   McB = mind controls body
>   CCB = causal closure of body-physical world

A trivial solution is that MdB is denied, mind and body are the same. However, we uphold MdB and maintain that mind is immaterial because mind is experience and its communication, which is indeed immaterial, abstract. Then the first statement is compatible with the third, while the second, immaterial-McB, is a known fallacy, that of mistaken concreteness.

Accepting MdB & CCB as our choice, we now face two difficulties: (1) If mind does not control body, what is it good for? (2) MdB may bar reduction: *Non*-reductive physicalism (NRP) holds that reduction is not possible if the explanandum is in another ontological class than the reduction base. Can mind (immaterial, abstract) nevertheless be reduced to neuronal (material, concrete) mechanisms? For (1) see Chapter 7, 'Consciousness'. For (2), see below.

6j. Mind-body problem

Around 1640 René Descartes famously proposed that bodily objects have physical extension (res extensa) while objects of the soul (or mind, res cogitans) do not. Yet they appear to interact [40, 41]. Since those days philosophers wondered how body and soul, extended objects and not-extended mind-phenomena, might influence each other [e.g. 84], which, according to 1st-person experience, they do. This is known as the mind-body problem of substance dualism.

In other words, if the mind is immaterial (an abstractum, without physical extension) while body and neurons have material properties, how can they encounter for physical interaction in space and time, how can mind affect body and body mind?

Suppose mind supervenes over neurons. And suppose there is a neuronal mechanism, a viewer, projector or decoder,[60] which operates on complexes C-A (of neuronal carrier C and abstract content A) by ignoring C and displaying A. Then the step from neuron to mind is

n → m: The neuronal gives rise to mind by performing the abstraction from neuronal carrier. The point is that the not-extended A is revealed by abstraction, but without A being changed in the process. Thus the not-extended A need not interact. For the viewing-process cannot affect the displayed content by changing A to A'. Instead, a neuronal mechanism may change C-A to C-A', the viewer now displaying A'.

The corresponding reverse step m → n is not possible. Abstract mind-experience cannot cause (ANA). The displayed content

---

60 Above, this ' display' is referred to as 'inner theatre stage'.

cannot affect the projector.

In conclusion, contrary to Descartes' famous speculation but in conformity with the principles ANA, CCP and STN, material body can reveal but cannot *and need not* affect abstract mind-content. Further, abstract mind-content cannot *and need not* affect material body. Rather, as shown in Figure 6.1, interaction is restricted to the neuronal level. The neuronal change C-A → C-A' is experienced as an illusionary mental A → A'.

6k. On reducing the mind (Red B)

The mind is taken to be a symbol for a bundle of experienced mental processes, including the psychological predicates. Each of these first-person predicates (like recalling, feeling, judging etc.) may be a system behaviour (SB), presumably generated by a special neuronal mechanism made of physical components. The mechanism is to be analysed in the third-person perspective. Then not the mind but each experience of a predicate-process is to be separately reduced to a neuronal process. For the reduction, the mental predicate-process and its suspected neuronal base should have the same causal role. Further, there should be *bottom-up* manipulability. Here is an example:

To recall a number is commonly seen as a mental operation. A number is an abstractum, immaterial, and 'recalling' is a function seemingly due to an immaterial mental process. When someone is given the mental task to recall his telephone number, specific neuronal processes will come into play. Perhaps like this: By physical interaction of neurons neuronal agents retrieve a pattern from long-term storage and make it conscious. Now the pattern can be experienced by the Self as the number in question. Since the neuronal world is subjectively unknown, the Self attributes the recalling to a (fictive) first-person mental process and not to the responsible third-person neuronal process.

Our task is to reduce the phenomenal 'mental process of recalling'. We treat the mental process as an explanandum under a 3rd-person perspective (Red B, Chapter 4). We investigate the brain, searching for a neuronal process of the same causal role. The process should precede or coincide with the conscious experience and under different circumstances always correlate with it. Further, its modification should modify the conscious experience and its interruption should abolish it. If these expectations of *bottom-up manipulability* hold, we have a candidate reductive mechanism explaining the mental phenomenon of 'recalling a number'.[61]

The different idioms associated with the levels provide a difficulty. At the mental level concepts like 'number', 'recall' and 'memory' are used. At the neuronal level we have 'activation', 'spike pattern', 'synaptic population response', 'long-term potentiation' etc.[62] Thus level-specific idioms make a reduction seem questionable. However:

> Not these descriptors and concepts but the functional identity at two levels and the *bottom-up* manipulability provide the clues leading to reduction of type Red A or B.

The Self-agent is experiencing the mental processes in a workspace, on a theatre stage.[63] This posting or display may allow the coordination of sub-agents. Thereby a function of the Self would be realized and should allow functional reduction via neuroscientific experimentation. In short, the first-person aspect of a Self-agent experiencing mental processes may possibly be functionally reduced to third-person neuronal mechanisms.

61. How to falsify

The so-called **"causal closure of the physical world"**, CCP,

---

61 Despite much research on number-phenomena and correlating brain events, to my knowledge this particular reduction was not yet attempted. But it seems entirely possible, using subjective reports with fMRI for localisation and methods like transcranial magnetic stimulation (TMS) for *bottom-up* manipulations.
62 Bennett and Hacker pointedly criticised the mixing of these idioms [13, 14]. However, in a bridging relation both must be used.
63 see Baars' model in Chapter 7, 'Consciousness'.

means that *every physical change which has a cause, is caused by physical interaction.* This dictum, based on countless observation, may be treated as a strong hypothesis. How can the hypothesis be falsified? One has to document a physical change which, without alternative, is caused by a non-physical phenomenon.

Often the 1st-person impression that our thoughts obviously can control our body, is taken to provide such a falsification, the thought being taken as immaterial, outside physics. But there are arguments which hold against this common notion:

(1) A thought, like all abstracta, has no causal power of interaction. This because it cannot encounter physical objects in space and time. Thus, to attribute causal power to an abstractum means to be trapped by the fallacy of mixing categories, a reification [139:51].

(2) There is an alternative, less dramatic interpretation: that a thought, though immaterial, has material roots. This neuronal basis, over which the thought supervenes, provides for a gap-less physical chain of causation, in which the immaterial thought seemingly participates.

Based on these arguments the causal closure hypothesis remains in-falsified by thought issues.

The related hypothesis of **reductive physicalism** states (my wording):

> Every material or immaterial phenomenon above a physical zero-level has a material physical base over which it supervenes and to which it may be reduced.

How can this dictum be falsified? One has to document a phenomenon which has no material base to which it may be reduced.

For an example, take numbers: "Two rocks" is concrete, the rocks are material, can interact. But "two" without "rocks" is due to an abstraction process, is immaterial. A number without the numbered cannot interact. Now the dictum says that a number, like all immaterial phenomena, has a material basis. This reduction base is found in the mechanism which performs the abstraction. That would be the abstracting module(s) of the human brain. Without them there would still be the "two rocks" but not the immaterial number "two". Numbers, like abstracta generally, have a neuron-dependent existence.

Note that numbers have Brentano-intentionality. Like parts of a communication they are *about* the (unspecified) numbered. As Franz Brentano famously observed, this *about* is the highlight of (I add: many) mental phenomena [21:124]. All abstracta are mental, are due to a neuronal abstraction or processing mechanism, many are *about* the abstracted, which, however, may not be specified. Thus we can restate our hypothesis with respect to mental phenomena:

> Immaterial phenomena are mental and have a material neuronal basis to which they may be reduced.

It is the neuronal abstracting or processing mechanism which provides the micro-phenomena to which the mental *explanandum* can be reduced.

The difficulty to falsify this hypothesis makes reductive physicalism a strong position. Why is it nevertheless often rejected? First, it contradicts our naïve or primary realism, which is subjectively blind to neuronal phenomena. If one adopts this realism as an unshakable position, physicalism cannot be accepted.[64]

Further, the reductions of physicalism, though largely feasible, were argued to leave an irreducible remainder, the qualia. First-person qualia-feelings like those resulting from tasting a wine, according to Kim, cannot be functionalized and, therefore, reduced with his 'functional strategy of reduction' [74]. However, as pointed out above (Section 4i), qualia-feelings do have a function by guiding our well-being [33:55]. Thus they can be functionally reduced, at least in principle. Contrary to Kim's view, qualia fail to falsify RP.

---

64 Also, regarding universal system levels, their basal events are exclusively physical, by axiom. Such an axiom may be rejected.

7. Consciousness

Consciousness is described as the neuronally founded ability to experience as a Self-agent. Proto-, phenomenal and access consciousness are distinguished and their features enumerated. Reduction to physical base by material synthesis and other strategies is explored.

In the framework of *perspective dualism* [65] consciousness is the third-person objective ability to be awake, aware and attentive, alert and comprehending and the first-person subjective experience of our inner theatre stage and through it of our surroundings. Thereby we can flexibly respond to changes and deal with novel situations by focussing our neuronal resources. There is a core-consciousness of the *now* (which some animals posses, too) and an extended, autobiographic consciousness [33], both supporting agent-world polarity. Ned Block made the similar and widely accepted distinction [19] between a phenomenal consciousness (having access to current outer-sense percepts and body sensations) and an extended access-consciousness (having in addition access to autobiographic memory and to sub-agents working unconsciously).

Despite all efforts and occasional claims of success [e.g. 39], the nature of consciousness remained a mystery for long. One

---

65 [21, 57, 58, 101]

difficulty was and is that research reveals many details but relies on subjective reports of content. Privileged access to content is given only to the conscious Self, the first-person. Its report alone can reveal meaning. But is the first-person real, is it independent from our thinking, independent from our neuronal activity? Apparently not.

Consciousness seems to be a major achievement of evolution, yet what is the 1st-person phenomenon 'good for' if all the action is 3rd-person neuronal? Further, 'who' is conscious? How does raw experience arise? That is David Chalmers' "hard problem" [25, 123]. It was suggested to temporarily ignore these difficulties of the first-person perspective and focus instead on neuronal correlates of consciousness (NCC) [32]. However, recent progress is significant [5] and the stakes are high for the ongoing research efforts attempting to solve "the hard problem".

7a. Global work-space

The principal EEG signature of consciousness is long known: a fast, irregular, low-voltage signal. This beta-activity appears to be due a to de-synchronisation of parallel cortical areas, perhaps their occupation with more detailed tasks and less distributed communication about them. The beta-activity is associated with a feeling to be awake and in control, to be a Self, a feeling, sensing, thinking autonomous first-person agent who can plan and act on its environment and remember its activity (autobiographic memory).

One conceptual highlight of research is the global workspace model of consciousness proposed by Bernard J. Baars and colleagues [6, 8, 102]. The work-space of consciousness is experienced as a display, a monitor screen or a mental simulation-stage. Wake-state-experience is a multimodal sketch on this "theatre of consciousness". The sketch displays live information to a group of neuronal agents (likely located in distinct cortical areas) represented by the Self (see below). The agents merely scan the displayed content which, being abstract, is not interactive itself. (It is the neuronal base of the content which mediates a causal relevance.) Thus a communication aspect of consciousness is defined, which fits nicely with Brentano's language-like *about*-ness of the mental [21].

The message, then, is cast in the idiom of a flow of multimodal theatre scenes which can be experienced. In this 1st-person picture-language the Self and specialised neuronal modules (agents represented by the Self) read or scan the multimodal overall situation in present, past and future:

1. Present. The ability of the Self, to experience in the 'now' (working memory) new sensory content in a multimodal scenic way, on-stage.
2. Past. The ability to remember such content and display it on-stage in a quasi-sensoric way.[66]
3. Future. The ability to invent fictive scenes and to explore their consequences (prediction).

Any involvement of the Self, any presentation to the conscious, takes time in the order of 0,3 to 1 seconds [115:133]. This pronounced time requirement points to a complex neuronal mechanism of presentation.

Why consciousness? Above it was argued that there is no *top-down* interactive effect, no causation arising from the mental level. The mental, being immaterial, cannot cause biophysical effects and the physical is causally closed. Further, when viewed from the mental level, the neuronal is transparent for lack of sensors to monitor its activity. The path neuronal - mental is *bottom-up* only. If so, is the conscious mental experience of content an epiphenomenon, without consequence? Why was it preserved and enhanced by evolution, why is it not sufficient to keep things unconscious?

With their 'global workspace' model Bernard J. Baars and colleagues gave an interesting partial answer to this question. Apparently, unconscious agents are not easily synchronised to jointly solve a new problem. Baars suggested that content ex-

---

66 While remembering a sensory experience the same cortex areas are active, which were also active during the primary sensory experience [5:53].

perienced by the conscious Self is for internal communication. Like on a bulletin-board or on a monitor-screen this content becomes available to the various brain modules acting as specialised neuronal agents. It allows such agents, which are partially autonomous, to scan the bulletin-content and coordinate their activity accordingly, working together on the solution of an onstanding, novel problem. In this model the physical root of conscious content has an important role of causal relevance in the neuronal processing of the brain [7, 8]. In their 'wagon wheel' scheme the neuronal modules involved are specified [102:1138].

Recent research has strengthened the communicative aspect, leading to Tononi's "integrated-information theory of consciousness" [9]. Accordingly, the loss of consciousness (LOC) in narcosis is associated with a disruption of communication among cortical modules [86].

Thus, the "bulletin-board" is not only mental but first of all neuronal, for, as pointed out with Figure 6.1, content itself does not interact, it is the brain modules which act by means of their neuronal causal chains. In terms of systems theory, the work of several neuronal agents is holon-bundled on level m-1, there is a *mechanism of concerted action* of modules which alone can solve a novel problem. And the Self represents these bundled agents on system level m, displaying their SB.

According to present insight, the neuronal basis of the conscious Self is the so-called executive system of the dorso-lateral prefrontal cortex and the anterior cingulate cortex [5:49,51]. It is remarkable that activation of these conscious-executive areas is greatly diminished when execution becomes automatic by repetition. Then activity is shifted to basal ganglia and cerebellum. Conscious problem solving is new-problem solving.

In summary, the conscious-executive cortical areas are involved in the pre-automatic solving of novel problems. Their activation is accompanied by the subjective mental experience of control and 1st-person agency, and of a multimodal sketch, displayed on the 'inner theatre stage', which can be remembered. The topic is under intense investigation.

7b. On reducing consciousness

In my wording, consciousness is foremost the ability to experience as a Self-agent. It is based on neuronal processes which give rise to (a) activation (b) evaluation and (c) insightful responses and cogitation.[67] Here is a summary:

(a) Proto-consciousness: An *activating* neuronal process enabling the organism to be awake, attentive and ready to act. The process is mediated by the ascending reticular activating system (ARAS) of the brain stem [33:248], which effects a pulsed release (4-40 Hz) of noradrenalin throughout cortex, thalamus and many other brain locations.

(b) Phenomenal consciousness: (Core-consciousness of the moment) The feeling Self attains agent-world polarity with access to outer senses and body senses. It experiences the current stream of outer-sense percepts and *evaluates* them by association with the synchronic body feelings, which report the degree of well-being. Thus the current multimodal scenes obtain a feeling quality. The evaluation can be reported to peers by behaviour, encouraging or warning them. A motivation may be derived from the evaluation and behaviour adjusted such that the situation improves.

(c) Access consciousness: The A-conscious Self has in addition access to sub-agents working unconsciously and to autobiographic memory. The evaluation with body-feelings will be stored in autobiographic memory together with its multimodal scene. Thereby the evaluation can be re-experienced with the scene.

A-consciousness is a major achievement of evolution. It has many facets, serving for a flexible, insightful handling of newly arising situations. This is summarized in the following table: [68]

---

67 The division into phenomenal and access consciousness is due to Ned Block [19], the characterisation of feelings as markers of body states is due to Antonio Damasio [33] while work of unconscious sub-agents coordinated by a common work-space is due to Bernard Baars [7].

68 Compare to the 5 'axioms' listed by Aleksander [1:34].

	Consciousness is the (neuronally founded) ability to experience as a Self-agent (person, Self) who...	
1	is awake, attentive and ready to act (ARAS),	pro-con
2	*experiences* himself as Self in polarity to the world,	Ph-con
3	*evaluates* the current sensory stream of multimodal percepts with the synchronic body-sense and reports the evaluation by behaviour,	Ph-con
4	*experiences* multimodal scenes on an inner theatre stage or movie screen (B.J. Baars' work-space).	Ph-con
5	A stage from which sub-agents pick information, co-ordinating their unconscious work and reacting to demands	A-con
6	with a spotlight or projector which feeds the stage with scenes, for which contents compete.	A-con
7	The Self can *re-experience* on-stage scenes of the past, which were stored in autobiographic memory,	A-con
8	can make *predictions* of future events by assembling fictive scenes and drawing conclusions from them,	A-con
9	responds to novel scenes in a not-stereotype way, using *insight* into causal relations (while well-known scenes are handled by previously learned, unconscious-stereotype reactions),	A-con
10	learns from insights, assembling relations into a *world-model*,	A-con
11	derives own *identity* from autobiographic memory.	A-con

Table 7.1

The table contains for A-consciousness a bundle of criteria, several of them dealing with the handling of novel and past scenes. The bundle is laced together as experiences of a conscious Self-agent. Based on the table, the question what access consciousness is good for can be answered. The highlight is *insight*:

7. Consciousness

> By coordinating neuronal sub-agents, A-consciousness (of humans) permits the solution of novel problems, also by insight into their causal structure. The solution is not of the lexical, stereotype if-then type, but synthetic by conscious understanding of laws and causes. Since novel insight is novel, it cannot be achieved by subconscious stereotype response systems, even though subconscious agents may contribute to the solution when coordinated by scanning the 'bulletin-board'.

Given this result, how may an explanatory reduction of processes of consciousness to neuronal base be performed? First of all, since reduction is always a 3rd-person endeavour, the items of Table 7.1 are to be taken as 3rd-person objective phenomena (Red B, Section 4b). Then reduction of several items of the bundle listed in Table 7.1, as well as reduction of the experiencing Self-agent, viewed from the 3rd-person perspective, may proceed.

By way of example, let us deal with item 3, which is the key to phenomenal consciousness. Furthermore, let us choose reduction by material synthesis, we want to create a conscious artefact. By now this is the goal of many efforts of engineering [e.g. 1, 42, 51, 67]. The artefact should be a robot which experiences P-consciousness. Will this be possible?

Phenomenal consciousness requires feeling. When are we prepared to concede that an organism feels? Since feelings are body-markers of well-being [33], I argue:

> An organism fulfils the 3rd-person criteria of feeling if it evaluates its situation by temporal association of outer-sense percepts with synchronic body-percepts, transferring the evaluation of well-being to the former.
>
> Typically the organism will communicate its evaluation to its peers by behaviour.
>
> The organism will derive a motivation from the evaluation and follow the motivation by behaving such that the situation improves.

The same expectation applies to our robot. Here body-feelings may be realised by sampling the actual state of components, their degree of physical "well-being", and evaluate the states, say, as a feel-index. This is stored in a memory stack (autobiographic memory) together with the scene resulting synchronic from current outer-sense percepts. Thus the scene, by temporal association of two sensory streams, is evaluated with respect to well-being. The motivation-behaviour part can then be solved by applied cybernetics.

The result is amazing, seemingly we have built a feeling robot. Then, knowing that the robot is made of physical components, we may have reduced feeling to a physical base. Or have we?

Sure, we do not know *what* the robot feels, perhaps in the way of qualia. And, sure, we do not know this for animals and humans either, an epistemic gap which provides little comfort. In the case of other humans our 'certainty' about what they feel is a guess based on the similarity of peers. In case of robots, such similarity is not given. We only know that the robot fulfils the formal 3rd-person criteria of feeling in the evaluation of its well-being, but we do not know whether the robot has a first-person, a Self-agent experiencing feeling.

Having reached this junction, I take occasion to quote a short passage by Heinrich Heine. In his 1834 treatise "History of German Religion and Philosophy" the poet relates the following episode:[69]

> *They tell the story that an English mechanicus, having created most ingenious machines, finally conceived to fabricate a human. This he did, his handicraft behaved just like a man, it carried even in its leather chest some kind of human feeling [...] It could report its sensations in articulate sounds; in short, it seemed a perfect gentleman. For a real human nothing was missing but a soul. This, alas, the English mechanic could not construe and his poor creature, once conscious of its shortcoming, [...] implored its maker by day and night: "Give me a soul!"*

---

69 [61:120]. Translation by B.L.

The construction of a robot is an undertaking in the 3rd-person perspective while feeling is a 1st-person experience. Because of perspective dualism we cannot be certain, operating from the physical perspective, that the robot's physics gives rise to a Self-agent capable of such 1st-person experience, even if the robot fulfils the formal 3rd-person criteria of feeling.

But suppose the robot were to report about its Self by behaviour, e.g. by recognizing itself in a mirror, showing agent-world polarity. Further, the robot may pass the imitation game "Turing test" [132], conversing convincingly like a human. If such criteria are met, reduction by material synthesis possibly was successful. Possibly but not certainly.

Thus, for the moment we fail in the attempt to reduce phenomenal consciousness with certainty, using material synthesis. We cannot be sure that a conscious, experiencing Self is present. But we would concede the possibility if the above criteria are fulfilled.

Fortunately, material synthesis is not our only strategy. It maybe interesting to turn to A-consciousness and *test for understanding*. Given a novel problem, it should be possible to decide whether a lexical if-then solution is attempted, or a not-stereotype solution based on an insight into laws and causality of the problem and leading to correct predictions. If insight and prediction is in evidence, points 8-10 of Table 7.1 are fulfilled, which are highlights of A-consciousness.

Changing strategy further, we can attempt *reduction to a neuronal mechanism* (Section 4b) of selected items of Table 7.1, or *functional* reduction as specified before ([74] and Section 4f). Such reduction, of course, is a matter of detailed experimental research. It is encouraging that all items of Table 7.1 are task-oriented, are functionally defined.

The conscious Self, too, has a function, it operates the multimodal display by experiencing content and it questions and coordinates the sub-agents, which it represents and which scan the display. Thus there is a chance that all items and the Self are piecewise reducible to neuronal processes, without remainder, as expected by reductive physicalism.

7c. Culture

Brains do not live in isolation, they interact with their natural environment and their human environment. The latter gave rise to a spectacular coalition of brains, since ancient times the basis of human culture.

Indeed, our conscious mental life is to a large extent shared with our peers by means of communicated concepts, often passed down to us over many generations. These culture-based concepts form our mind, the richness of which is the central characteristic of our species.

Concepts, of course, are abstracts, they are 'about', are characteristic products of the human mind, which supervenes over our neuronal system.

Many concepts, though contingent, are strongly upheld by a group of people, supporting group-coherence, group-identity. A stranger is one with foreign concepts. Integration means to accept concepts of another group.

The many concepts which we have accepted are interdependent and together form a precarious structure which was built with difficulty and which is valuable to us as our "Weltanschauung" (world-view). We promote it, campaign for it in many conversations and defend it as part of our identity, even fight for it physically, with much emotional by-play.

C.P. Snow has observed that people prefer one of "The two cultures" [128], either the art-humanities or the science-engineering variety. Across the mental divide "they don't talk to each other" and, by excluding the other side, "they don't know what they are missing". Snow's ideal is a generalist, at home in both worlds, perhaps like the Renaissance artist and scientist-engineer Leonardo da Vinci. Notably, in terms of perspective dualism Snow's art-humanities culture is dominated by 1st-person experience while his science-engineering culture is a 3rd-person effort. Snow's people live perspective dualism.

Human culture is the characteristic achievement of Homo sapiens. It grows and changes much more rapidly than our genome. Culture has given rise to (by now planet-wide) endeavours like arts, sciences, engineering, philosophy, religion-ethics, sports etc., where every competent mind is invited to participate in the 'coalition of brains'. Still, change often seems not fast enough. True progress, like true knowledge, is slow.

> *"Knowledge is slow afoot,*
> *but wisdom limps far behind".*[70]

Cultural progress may sometimes be difficult to recognise. It causes collateral damage and some conservative minds reject it altogether. Yet, in a complex world bordering at deterministic chaos cultural change, like change generally, cannot be avoided. Where will it lead us?

---

70 Carl J. Wiggers, Handbook of Physiology, Sect. 2, Volume I, 1962.

8. *Summary*

Initially we asked whether the common intuition "humans are more than physics" is justified. Are humans more than neuronal machines, assemblies of neuronal mechanisms? Since we have come a long way.

2. Physical mechanisms were defined. Designed by man or evolved in nature, they are physical devices made of interacting components, optimized to alter the environment in a characteristic and quantitatively more or less predictable way. All models of mechanisms are causal models. Their deterministic and probabilistic versions were exemplified.

Turning to the descriptive levels of physics (Section 2b), it was found that the deterministic models of the macroscopic level are idealisations (IDM). The more general and more realistic versions are probabilistic models of the mesoscopic level. However, their stochastics are confined by the rules of the state-transition-diagram, STD, and leave no gaps in the physical causal chains. This is also true for neurons, which are constituted of molecules with stochastic behaviour.

STDs of mechanisms with steady state performance are cyclic. Their system behaviour SB shows the over-sum effect, it is more than what is expected from the uncoordinated components (OSE). This 'more' is due to the organisation. It does not preclude a reduction of SB to the mechanism and further to physical base.

3. Systems theory was introduced as a means to order phenomena of world and mind. Its levels were found to be pragmatic conveniences: by referring to mechanisms (holons) with symbols on the level above, we drastically reduce the number of items to be considered at once. However, the popular linkage of levels, either causally or by constitution, is problematic.

For parts-whole systems the possibility of *top-down* changes is to be denied all-together. If parts constitute a whole *exclusively*, then the whole can be changed *bottom-up* only, by changing the parts. It is not possible to change the whole in a *top-top* way and then observe *top-down* effects on the parts. Any *top-down* influence from whole to parts, though an ancient holistic concept, is impossible (TDD).

A novel concept, that of **universal** system levels, was developed. The levels contain the same basic physical elements, each level is all-comprising. Thus basic events are the same, occur synchronic on each level. The level-specific objects and events are symbolic. They arise from different grouping of the basic elements and assignment of symbols to the groups. Context and idiom are also level-specific. There is no need for a further coupling of levels at all. Coupling is already complete by basic identity, each level being basically the same. Since basal events are physical by axiom (RP-hypothesis), the physical basis of all complex phenomena is implied (USL).

4. Reduction or naturalisation is the attempt to understand an explanandum or macro-phenomenon from micro-processes and their laws at a lower system level. Where we cannot reduce, we cannot understand. Reduction is best done in a single-level step down to the next-lower level. When extending the range by skipping intermediate levels, explanations become less convincing or even incomprehensible (PLC).

Based on the 1st-person perspective alone, reduction of the mental to the neuronal is impossible, the neuronal being subjectively transparent (STN). Yet a 1st-person experience may be reduced in a 3rd-person perspective (Red B, Table 4.1). Several strategies for reduction in a 3rd-person perspective (Red A) were delineated: reduction to a model, to a known mechanism,

8. Summary

by material- or by simulated synthesis and functional reduction. In any case, to reduce a system behaviour at level n to a mechanism at level n-1 (single-level reduction) we make use of recursive *bottom-up* procedures.

Life was viewed as the over-sum system behaviour based on a bundle of interdependent sub-processes. Life may be reduced to those sub-processes and these may be separately reduced to physical events, for instance by material synthesis or by simulative synthesis. However, such a multi-level reduction involving many steps may not lead to a subjectively convincing explanation of life in physical terms. Then a feeling will remain that life is more than physics, with agency and autopoiesis it seems on a higher organisational level than physics. Indeed, this is the case: Life as over-sum behaviour is more than the sub-processes and their relations. Yet it may be reducible to physical base without remainder.

When discussing functional reduction of the **mental** in a general way, Jaegwon Kim argued that the much-discussed **qualia**, the subjective quality-properties of our raw feeling-experience, are without function. For him they constitute the irreducible residuum of an otherwise functionally reducible mental world. I hold, however, that qualia are task-oriented in the sense of Damasio's markers of body states. By their temporal association with synchronic outer-sense percepts the markers help to avoid the unpleasant, the unbecoming scene and seek the pleasant, becoming scene. In this sense qualia are functionally reducible, they are not an unresolved residuum of the otherwise successful reduction of the mental.

5. **Neuronal systems** were viewed in the third-person perspective. In neuronal **coding** only the quantitative information (*how much?*) is represented by spikes, by a rapidly variable pattern of time intervals separated by nerve pulses. The contextual part of the information (*what* and *where?*), I argue, is based on the identity of the neuron, mediated by surface markers and connectivity (MMC).

A **hierarchy** of functional models has been developed, spanning from molecules to neuronal networks, brain modules and brains.

On the molecular level the model performance is probabilistic or, in large ensembles, quasi-deterministic. At higher levels of organisation it may be nearly but not fully deterministic. For, in view of the molecular components, a residual degree of uncertainty cannot be excluded.

Reliability of mechanisms is apparently instrumented. The scale from fully random to quasi-deterministic provides an additional dimension for coding, low signal strength being associated with randomness and high strength with predictability. Population averaging across many parallel neurons will improve the overall reliability of a neuronal mechanism. That may be one reason why the human brain contains so many billions of neurons.

In the new concept of **universal** system levels, basal events occur synchronic on each level but are noted as changes of different higher-order neuronal structures. The idiom describing these structures and their relations necessarily differs from the idiom describing the basal events. It were highly impractical to specify, say, a spike in terms of basal events. Therefore a series of translation-levels is inserted above basal level for epistemic reasons. After all, our working memory can handle few items at once (USL, PLC).

6. The Mind was viewed in two perspectives. In the *first-person perspective* the mental (the mind) is the domain of a Self-agent who has the conscious *experience* of mental processes. These include the 'psychological predicates', thought about communicated mental content and causal power, the subjective experience of control and the experience of a multimodal sketch, displayed on the 'inner theatre stage', which can be stored in autobiographic memory.

The Self-agent views itself to exist apart from the material world in an agent-world polarity. It views itself as related to its body but is not aware of its own neuronal system. This comes about because we perceive only what our senses tell us, and we have no sensory organ to notice our own neuronal activity. Thus arises the illusion of an autonomous Self separate from the material world but acting on it, the actions apparently driven by immaterial thoughts.

8. Summary

Can mind have mechanisms? Suppose somebody is thinking and writes his thoughts down on paper. (a) There is a *causal chain* of neuronal, biophysical and physical interactions which, without gaps, constitutes this activity. We may call this activity *bottom-bottom*.

(b) The thinker has an experience, he becomes conscious of the content and its change, of a flow of content. The flow is due to a sequence of discrete *bottom-up* events which raise content into consciousness. However, since the bottom dimension is transparent (STN), it will appear as a *top-top* flow of content. Clearly, the *bottom-bottom* biophysical events have causal power, but the abstract content does not. There is no way down, nor is it needed: whatever is to be relayed down, is already there.

Mind cannot participate in physical interaction. For mind is immaterial, is conscious experience *about*. Experience, being immaterial, must be causally powerless (ANA). Rather, being *'about'*, it serves for communication.

Accordingly, the Bieri-trilemma is solved by rejecting 'immaterial mind controls body' as a known fallacy, that of mistaken concreteness. In this vein, the long-standing mind-body problem finds its solution: there is no need for the impossible physical interaction of immaterial mind and physical body, the causal power of neuronal *bottom-bottom* events suffices.

In the *physical perspective* of the mind, conscious-executive cortical areas serve for the pre-automatic solving of novel problems. They act as sub-agents which communicate, coordinating their unconscious efforts. Their activation is accompanied by the subjective mental experience of control and Self-agency.

Reduction of mind: The Self-agent is experiencing the mental phenomena in a work-space, on a theatre stage. This 'posting' allows the coordination of sub-agents. Thereby at least one function of the Self is defined and should allow functional reduction.

The mental predicate-processes are to be reduced separately to neuronal processes. In each case, the mental predicate-process and its neuronal base should have the same causal role. Further, there should be *bottom-up* manipulability. Then the first-person aspect of agency experiencing mental processes may by future work be reduced to third-person neuronal mechanisms.

Further, concepts of psychology and even sociology will have a physical base to which they relate indirectly by a chain of symbols. However, a long chain of reductions obscures explanations. Level-specific idioms obscure explanations too, make them incomprehensible. Yet, the principle of properly grounded *universal levels* implies reduction of all immaterial phenomena to physical mechanisms and, in the end, basal events.

7. Consciousness is the third-person objective ability to be awake, attentive and comprehending and the first-person subjective experience of an inner theatre stage (multimodal scene) and through it of our surroundings. Consciousness is based on neuronal processes comprising (a) activation of cortical facilities (b) evaluation of the current sensory percepts and resulting scenes and (c) the neuronal equivalent of cogitation, prediction and flexible insightful responses.

Recent research has strengthened the communicative aspect of consciousness, leading to Tononi's 'integrated-information theory'. Already with their 'global workspace' model Bernard J. Baars and colleagues gave a partial answer to the question, why consciousness is of advantage. It serves for internal communication. Like on a bulletin-board or on a monitor-screen conscious content becomes available to the various brain mechanisms acting as specialised neuronal agents. It allows such sub-agents to work together on the solution of on-standing, novel problems.

The primary conscious-executive brain modules were identified by experimental research.

In *phenomenal consciousness* the Self has access to outer senses and body senses. It experiences the current stream of outer--sense percepts and *evaluates* them by association with the synchronic body feelings, which report the degree of well-being with qualia. In *access-consciousness* the Self has in addition access to biographical memory, which defines the Self's identity,

8. Summary

and to neuronal sub-agents working unconsciously. Access-consciousness serves for a flexible, insightful handling of novel situations.

Several strategies were considered to reduce processes of consciousness to neuronal mechanisms. They show feasibility, but the detailed reduction itself will require considerable efforts in neuroscientific research.

During the course of this treatise the following rules and principles were encountered:

IDM: Idealisation in deterministic models (Sections 2b, 2d). In the neuronal system, quasi-deterministic and deterministic models are idealisations focussing on the mean behaviour of an ensemble of N units, neglecting fluctuations. The D-models of level-3 (Sect. 2b) are limiting cases of more realistic probabilistic models of level-2. Yet fluctuations, though not thematized in D-models, do not vanish but even increase with ensemble size N. Fluctuations remain irrespective of our models, the prediction of means can be misleading, full reliability of model predictions is not possible.

OSE: Over-sum effect (Section 2f). Steady-state models are based on cyclic STDs. This organisation of interactive components allows for 'over-summative' behaviour (m-causation). Due to the over-sum effect the SB-property of a whole can be 'more' than component behaviour but reducible to (fully explainable by) the interactive component-level. While reduction may extend down to physical base, comprehension hardly follows, due to the PLC.

TDD: *Top-down* denied (Section 3c). For parts-whole systems the possibility of *top-down* changes is to be denied all-together. If parts constitute a whole *exclusively*, then the whole can be changed only in a *bottom-up* process, by changing the parts. It is not possible to change the whole in a *top-top* way and then observe *top-down* effects on the parts. Any *top-down* influence from whole to parts, though an ancient holistic concept, is impossible.

USL: Principle of universal system levels (Section 3f). Viewed in the 3rd-person perspective the levels contain the same basic physical elements, each level is all-comprising. Thus basic events are the same, occur synchronic on each level. The level-specific objects are symbols assigned to different groups of the basic elements, necessitating level-specific context and idiom. There is no need for a further coupling of levels at all. Coupling is already complete, each level being basically the same. Since basal events are physical by axiom (RP-hypothesis), the physical basis of all complex phenomena is implied.

MMC: Myth of the missing context (Section 5a). The fixed or contextual part of neuronal data should answer questions about the *what* and *where*. Such not-graded information can hardly be represented by pulse patterns. I suggest that the contextual part for instance of the primary sensory message is based on the identity of the primary neuron. Thus information about w*hat* and *where* is already contained in properties of the sensory neuron. It is due to matching connectivity to specificity in an autopoietic process. This process seems to realize part of the construction plan of our neuronal system.

PLC: Principle of limited cognition (Sections 5c). The brain's abilities are large but necessarily limited. For instance, the working memory of humans can handle only 10 or less items at once (Miller's number [5:259, 96]). We therefore create upper system levels, where multiple components are represented by one symbol, such that fewer items need to be considered. Multi-level reductions have little subjective explanatory value because intermediate explanations are skipped, necessitating the handling of too many items.

STN: Subjective transparency of the neuronal (Section 6a), with 'transparent' taken in the sense of 'present but unnoticed', like glass to vision. The activity of our neuronal system is subjectively transparent because we have no sensory organ for such activity [e.g. 87:240, 95:188,289, 97, 104:267, 105:95]. 'Through' the unnoticed neuronal system we sense features of world and body. And we do not miss what we cannot sense. This helps to establish the notion of a mind and Self existing independently from the neuronal-physical world.

ANA: Abstracta non agunt in concreto (Section 6c). Abstracts are not locatable in physical space, they cannot interact physically, are without causal power. To attribute causal power to an abstractum means to make a category-mistake, the fallacy of mistaken concreteness [139:51]. Thus abstract thoughts (immaterial thoughts) cannot cause.

CCP: So-called "causal closure of the physical world" (Section 6e). The expectation, based on experience, that every physical change having a cause is caused by physical phenomena. Non-

8. Summary

physical entities are necessarily immaterial, therefore do not interact with material objects, cannot encounter them in space and time. In consequence, an immaterial mental phenomenon alone (like a thought) cannot 'cause' physical changes at neurons.

Among these rules we find two general restrictions of our brain function:

1. The sensory blindness to the neuronal (STN). It supports the illusionary independence of the inner perspective from the neuronal and strengthens the "more than physics" intuition.

2. The limited cognitive capacity (PLC). It forces us to break complexity down into palatable chunks and makes multi-level reduction incomprehensible even though it is formally possible.

Based on such limitations humans do experience Self-independence and feel to be more than physics (and rightly so, in the sense of the over-sum effect, OSE). Yet, contra-intuitively, the human over-sum faculties are reducible to physical base by multi-level reduction (as the OSE example of mechanisms shows). However, in case of such multi-level reduction comprehension tends to fail (due to PLC), re-enforcing the "more than physics" stance. The arguments will be concluded in a *facit* below.

9. Facit

Reductive physicalism:

We parse world phenomena into system levels to make them epistemically palatable, partly because processing in our working memory is limited to few items at once [96]. Single-level reduction means to translate from the order of one system level to the order of the level below. RP claims per hypothesis that this is always possible.

Reductive physicalism cannot be proven. I have presented it as a hypothesis, often rejected [e.g. 24, 69] but not formally falsified. In my wording, its general form, including the mental, runs like this:

> Every phenomenon above zero-level, immaterial and material, has a physical base to which it may be reduced.

In case of mental phenomena this basis is found in the processing of neuronal modules, which gives rise to conscious experience of thought and other immaterial, abstract explananda (Red B in Section 4b).

The hypothesis of reductive physicalism rests on the causal closure hypothesis of physics (CCP). The mind cannot participate in physical interaction. For mind is immaterial, is conscious experience *about* its topoi. Experience, being immaterial, must be causally powerless. Rather, by being *'about'*, it serves for

communication. Language, description, internal communication of brain modules is a key-feature of consciousness.

Yet the mental, seat of the conscious and without causal power, seems to have causal relevance. I consciously deliberate my reasons which then appear to guide or even cause my action. This guiding, however, is due to the material roots of such immaterial conscious deliberation, part of the neuronal-physical world. The roots provide a backbone of causal interactions (Figure 6.1), a backbone which is revealed by our reductions as we aim to understand world and mind.

Post-reductionism:

In brief, post-reductionists argue that single-level reduction of the mental to neuronal base and its multi-level reduction to physical base is not possible because (a) the reduction base is incomplete and (b) the mental is more than a neuronal or physical system.

(a) is true (Sections 2b, 4j). Yet reduction base may be completed by further empirical work and is already complete enough for many Red A and some Red B reductions.

(b) is an intuition from the first-person perspective. It seems acceptable but does not necessitate failure of reduction. For the mental may be over-summative, *more* than its one-level reduction-base and yet reducible to it and to physical base without remainder. This would be the case when 'more than' refers to a whole symbolizing components and their interaction. An over-summative whole owning the SB is always 'more' than its components and their interaction, but explained by their interaction, it is reducible. Thus a general refutation of reductions does not follow from the arguments of post-reductionism. Fortunately: for who wants to explain, needs to reduce.

Based on these observations we may now draw a conclusion.

9. Facit

Conclusion:

The neuronal system has biophysical *mechanisms*. Their models, if deterministic, were found to be idealisations of more realistic probabilistic models. It was particularly noted that a whole with the system behaviour of a mechanism will show over-summative behaviour (OSE), i.e. more than what is explainable by individual components, yet fully explainable by physical laws describing their concerted action. Thus, if the mental were the SB of several neuronal modules, it would be at once 'more' than these and fully explainable by their interaction.

The construction of *system levels* was suggested to follow epistemic reasons. We zoom out to higher levels, where less items need to be kept in working memory (PLC). The novel concept of *universal levels* (USL) specifies basic events which occur synchronic on each level. The level-specific objects arise from different hierarchical grouping of identical basic elements. There is no need for a further coupling of levels at all.

Looking at *reductions* in detail, it was found that single-level versions have clear explanatory value while multi-level reductions, which skip intermediate explanations, are less comprehensible (Section 5c). This may enhance a 'more than...' intuition following a multi-level reduction towards physical base.

A number of reduction strategies were discussed and tentatively applied to various topics including life, mind, art, agency, dignity and consciousness (Sections 4d, 6k, 4h, 4d, 6g, 7b). A principal feasibility was found. Of course, to actually perform such reductions will require detailed neuro-experimental work.

Judging by feasibility, then, first-person mental phenomena may possibly be understood in terms of neuronal mechanisms of the third-person perspective, using single-level reduction based on experimental work. Multi-level reductions to physical base, while also possible, tend to be of little explanatory value. Here the mental may be reducible to physical base but subjectively it will not be convincingly explained by physics because such explanation over-taxes our cognitive abilities (PLC).

Perspective dualism originates from an illusionary 1st-person perspective: the inside view of a Self-agent claiming to exist separate from world and body, and acting and controlling by thought. This view is made possible by the subjective transparency of the neuronal (STN). The transparency hides the neuronal mechanisms responsible for our thinking and actions.

The *mind-body problem* was found to vanish: Immaterial mind cannot and *need not* interact with material body, the neuronal system already contains all the information to be transmitted by such interaction.

The initial question asked was, if and how the mental may be understood in terms of neuroscience and its mechanisms. The intuition claimed that this is not possible because human minds, or humans in general, are 'more' than what is explainable by physics. Humans have design, agency, values, dignity. Such phenomena of the first-person perspective have to be accounted for by eventually reducing the mind to neuronal and physical base. While this may be possible in scientific terms, it is a multilevel undertaking, subjectively of little explanatory value. The failure to comprehend will support the initial intuition of 'more than' physics. But, notably, an over-summative 'more than physics' is compatible with a multilevel reduction to physical base.

Such conclusions may modify our concepts of human nature. They stress and in no way belittle the 'more than physics' nature of the mind. Its features and concepts remain indispensable for our conscious life. After all the richness of the mental is the central characteristic of Homo sapiens.

Reductive physicalism as well as 'more than physics':

In brief, the 'humans are more than physics' intuition can be justified. It may be accounted for already by the described 'over-summative' system behaviour of physical mechanisms (OSE). Contrary to intuition, however, such system behaviour is quite compatible with reduction, even with a multi-level reduction, to physical base, as OSE shows. Yet such objective multi-level reduction, which skips intermediate explanations, is subjectively hard to comprehend, given our cognitive limitations (PLC). This tends to enhance the initial (and justified) 'more than physics' intuition.

9. Facit

10. Bibliography

1. Aleksander, I., *The world in my mind, my mind in the world*. 2007, Exeter: Imprint Academic. 196 pages.
2. Andersen, P.B., et al., eds. *Downward Causation. Mind, Bodies and Matter*. 2000, Aarhus University Press: Aarhus. 354 pages.
3. Aristoteles, *Metaphysik VII 10, 1041 b*. ca. 330 B.C.
4. Ast, F., *Grundlinien der Grammatik, Hermeneutik und Kritik*. 1808, Landshut.
5. Baars, B. and N.M. Gage, *Cognition, Brain and Consciousness: An Introduction to Cognitive Neuroscience*. 2007, London: Elsevier/Academic Press. 546 pages.
6. Baars, B.J., *In the theatre of consciousness: global workspace theory, a rigorous scientific theory of consciousness*. Journal of Consciousness Studies, 1997. **6**(2-3): p. 292-309, comments 310-364.
7. Baars, B.J., *In the theatre of consciousness: The workspace of the mind*. 1997, Oxford: Oxford University Press. 193.
8. Baars, B.J., *The conscious access hypothesis: origins and recent evidence*. Trends in Cognitive Sciences, 2002. **6**(1): p. 47-52.
9. Balduzzi, D. and G. Tononi, *Qualia: the geometry of integrated information*. PLoS Computational Biology, 2009. **5**(8, e1000462): p. 1-24.
10. Bechtel, W., *Mental Mechanisms. Philosophical perspectives on Cognitive Neuroscience*. 2008, New York: Routledge. 308 pages.
11. Beckermann, A., *Analytische Einführung in die Philosophie des Geistes*. 2. Auflage ed. 2001: de Gruyter. 498 Seiten.
12. Bennett, K., *Exclusion again*, in *Being Reduced*, J. Hohwy and J. Kallestrup, Editors. 2008, Oxford University Press: Oxford. p. 281-305.
13. Bennett, M.R., et al., *Neuroscience and Philosophy. Brain, Mind and Language*. 2007, N.Y.: Columbia University Press. 216 pages.

14. Bennett, M.R. and P.M.S. Hacker, *Philosophical Foundations of Neuroscience*. 2003, Berlin: Blackwell Publishing. 461 pages.
15. Berlinski, D., *On Systems Analysis: An essay concerning the limitations of some mathematical methods in the social, political, and biological sciences*. 1976, Cambridge, Ma: The MIT Press. 186 pages.
16. Bertalanffy, L.v., *Robots, men and minds. Psychology in the modern world*. 1967, New York: George Braziller. 150 pages.
17. Bertalanffy, L.v., *General System Theory*. 1969, New York: George Braziller. 296 pages.
18. Bieri, P., ed. *Analytische Philosophie des Geistes*. Neue Wissenschaftliche Bibliothek. 1993, Bodenheim. 372 pages.
19. Block, N., *On a confusion about a function of consciousness*, in *The Nature of Consciousness: Philosophical Debates*, N. Block, O. Flanagan, and G. Guzeldere, Editors. 1998, MIT-Press. p. 375-415.
20. Block, N., *Do causal powers drain away?* Philosophy and Phenomenological Research, 2003. **67**: p. 133-150.
21. Brentano, F., *Psychologie vom empirischen Standpunkt*. 1874, Leipzig: Meiner Verlag 1924.
22. Bunge, M.A. and M. Mahner, *Über die Natur der Dinge*. 2004, Stuttgart: Hirzel. 273 Seiten.
23. Cahan, D., ed. *Hermann von Helmholtz and the Foundations of Nineteenth-Century Science*. 1993, University of California Press: Berkeley, Los Angeles, London. 666 pages.
24. Cartwright, N., *The Dappled World: A Study of the Boundaries of Science*. 1999, Cambridge, UK: Cambridge University Press. 247 pages.
25. Chalmers, D.J., *Facing Up to the Problem of Consciousness*. Journal of Consciousness Studies, 1995. **2**(3): p. 200-219.
26. Churchland, P.M., *Eliminative Materialism and the Propositional Attitudes*. Journal of Philosophy, 1981. **78**: p. 67-90.
27. Churchland, P.S., *Neurophilosophy: Towards a unified science of the mind-brain*. 1986, Cambridge, MA: MIT Press. 546 pages.
28. Churchland, P.S. and T.J. Sejnowski, *The computational brain*. 1992, Cambridge, MA: MIT Press. 544 pages.
29. Clarke, P.G.H., *The limits of brain determinacy*. Proc. R. Soc. B 2012. **279**(1734): p. 1665-1674.
30. Craver, C.F., *Explaining the brain. Mechanisms and the mosaic unity of neuroscience* 2007, New York: Oxford University Press. 308 pages.
31. Craver, C.F. and W. Bechtel, *Top-down causation without top-down causes*. Biology and Philosophy, 2007. **22**: p. 547-563.
32. Crick, F. and C. Koch, *A framework for consciousness*. Nat Neurosci, 2003. **6**(2): p. 119-26.

33. Damasio, A.R., *The feeling of what happens. Body and emotion in the making of consciousness.* 1999, New York: Harcourt Brace & Co. 386 pages.
34. Danaher, J. *Top-down-causation.* philosophical disquisitions 2010 [cited 2013 Jan. 14th]; Available from: http://philosophicaldisquisitions.blogspot.de/2010/01/top-down-causation-part-1-mechanisms.html http://philosophicaldisquisitions.blogspot.de/2010/01/top-down-causation-part-2-constitutive.html
35. Davidson, D., *Mental Events*, in *Essays on Actions and Events.* 1970, Clarendon Press: Oxford.
36. Dawkins, R., *The blind watchmaker.* 1986, Oxford: Oxford University Press.
37. DeFelice, L.J., *Introduction to Membrane Noise* 1981, New York: Plenum Press. 500.
38. Dennett, D., *Real patterns.* Journal of Philosophy, 1991. **87**: p. 27-51.
39. Dennett, D.C., *Consciousness explained.* 1991, Boston, MA: Little Brown.
40. Descartes, R., *Discours de la méthode pour bien conduire sa raison et chercher la vérité dans les sciences.* 1637, New York: Prometheus Books. 123 pages.
41. Descartes, R., *Meditationes de prima philosophia.* 1901 ed. 1641, Paris.
42. Edelman, G.M. and G. Tononi, *A Universe of Consciousness. How matter becomes imagination.* 2000, New York: Basic Books. 274 pages.
43. Elias, P., *A note on the misuse of "digital" in neurophysiology*, in *Sensory Communication.* 1961, MIT Press: Cambridge, Mass. p. 794-795.
44. Eyring, H., R. Lumry, and J.W. Woodbury, *Some Applications of Modern Rate Theory to Physiological Systems.* Rec. Chem. Prog., 1949. **10**: p. 100-112.
45. Faisal, A.A., L.P.J. Selen, and D.M. Wolpert, *Noise in the nervous system.* Nat Rev Neurosci, 2008. **9**(4): p. 292-303.
46. Falkenburg, B., *Mythos Determinismus: Wieviel erklärt uns die Hirnforschung?* 2012, Heidelberg: Springer. 458 pages.
47. Falkenburg, B., *Wieviel erklärt uns die Hirnforschung?* Information Philosophie, 2012. **40**(1): p. 8-19.
48. Feinstein, P. and P. Mombaerts, *A Contextual Model for Axonal Sorting into Glomeruli in the Mouse Olfactory System.* Cell, 2004. **117**(8): p. 817-831.
49. Foerster, H.v., *Sicht und Einsicht. Versuche zu einer operativen Erkenntnistheorie.* 1985, Heidelberg: Carl-Auer 1999.
50. Folkow, B., *Increasing importance of integrative Physiology in the era of molecular biology.* News in Physiological Sciences, 1994. **9**(4): p. 93-95.

10. Bibliography

51. Franklin, S., *IDA, a conscious artefact?* Journal of Consciousness Studies, 2003. **10**(4-5): p. 47-66.
52. Frisch, M., *Kausalität in der Physik*, in *Philosophie in der Physik*, M. Esfeld, Editor. 2012, Suhrkamp: Berlin. p. 411-426.
53. Gazzaniga, M.S., *Who's in charge?* . 2011, New York: HarperCollins 260.
54. Gloy, K., *Die Geschichte des ganzheitlichen Denkens*. 1996, Köln, München: Komet Verlag & C.H, Beck. 288 Seiten.
55. Griffiths, D., *Introduction to electrodynamics*. 1989, New Jersey: Englewood Cliffs.
56. Guttenplan, S., *Reduction*, in *A Companion to the Philosophy of Mind. Blackwell Companions to Philosophy*. 1994, Blackwell. p. 535-536.
57. Habermas, J., *Freiheit und Determinismus* Deutsche Zeitschrift für Philosophie, 2004. **06/2004**: p. 871-890.
58. Habermas, J., *Um uns als Selbsttäuscher zu entlarven, bedarf es mehr*, in *Frankfurter Allgemeine Zeitung*. 2004. p. 35.
59. Hebb, D.O., *The Organization of Behavior. A Neuropsychological Theory*. 1949, New York: John Wiley & Sons.
60. Heil, J., *From an ontological point of view* 2003, Oxford: Clarendon Press. 268 pages.
61. Heine, H., *Zur Geschichte der Religion und Philosophie in Deutschland*. Heinrich Heine Werke. Vol. 4 Schriften über Deutschland. 1834 / 1968, Frankfurt am Main: Insel Verlag.
62. Hess, B., *The glycolytic oscillator.* J. exp. Biol., 1979. **81**: p. 7-14.
63. Hill, T.L., *Free Energy Transduction in Biology. The steady-state kinetic and thermodynamic formalism*. 1977, New York: Academic Press. 229 pages.
64. Hobbes, T., *De Corpore*, in *Elementa Philosophiae (Trilogie)*, K. Schuhmann, Editor. 1655, Meiner: Hamburg 1997
65. Hodgkin, A.L. and A.F. Huxley, *A quantitative description of membrane current and its application to conduction and excitation in nerve.* J. Physiol., 1952. **117**: p. 500-544.
66. Hohwy, J. and J. Kallestrup, eds. *Being Reduced. New Essays on Reduction, Explanation, and Causation*. 2008, Oxford University Press: Oxford. 312 pages.
67. Holland, O., ed. *Machine Consciousness*. 2003, Imprint Academic: Exeter.
68. Hong, W., et al., *Leucine-rich repeat transmembrane proteins instruct discrete dendrite targeting in an olfactory map.* Nature Neuroscience, 2009. **12**: p. 1542-1550.
69. Horst, S., *Beyond Reduction. Philosophy of Mind and Post-Reductionist Philosophy of Science*. 2007, Oxford: Oxford UP. 230 pages.

70. Hume, D., *An enquiry concerning human understanding*. 1741, Oxford.

71. Hüttemann, A., *Eine dispositionale Theorie der Kausalität*, in *Lebenswelt und Wissenschaft. Deutsches Jahrbuch Philosophie 2*, C.F. Gethmann, Editor. 2010, Meiner: Hamburg. p. 451-467.

72. Juarrero, A., *Dynamics in Action. Intentional behavior as a complex system*. 2002, Cambridge, MA: MIT Press. 288 pages.

73. Kim, J., *Blocking Causal Drainage and Other Maintenance Chores with Mental Causation*. Philosophy and Phenomenological Research, 2003. **LXVII**: p. 151-176.

74. Kim, J., *Physicalism, or something near enough*. 2005, Princeton: Princeton University Press. 186 pages.

75. Kistler, M., *Mechanisms and downward causation*. Philosophical Psychology, 2009. **22**(5): p. 595-609.

76. Klir, G.J., *Facets of Systems Science*. 2 ed. IFSR International Series on Systems Science and Engineering, ed. G.J. Klir. Vol. 15. 2001, New York: Kluwer Academic, Plenum Publishers. 740 pages.

77. Knorr-Cetina, K., *Die Fabrikation von Erkenntnis. Zur Anthropologie der Naturwissenschaft*. 1984, Frankfurt: Suhrkamp. 357 pages.

78. Koch, C., *Biophysics of Computation: Information processing in single neurons*. 1998: Oxford University Press. 552.

79. Koch, C., *Consciousness. Confessions of a romantic reductionist*. 2012, Cambridge, Ma: MIT Press. 182.

80. Koch, C. and I. Segev, eds. *Methods in Neuronal Modeling: From Synapses to Networks*. 1989, The MIT Press: Cambridge, Massachusetts. 523 pages.

81. Koestler, A., *The ghost in the machine*. 1967, London: Hutchinson 1967 or Arkana Penguin Books 1989. 384 pages.

82. Korn, H. and D.S. Faber, *Quantal analysis and synaptic efficacy in the CNS*. Trends Neurosci., 1991. **14**: p. 439-445.

83. Kuhn, T.S., *The structure of scientific revolutions*. 1996, Chicago: University of Chicago Press. 212 pages.

84. Lauth, B., *Descartes im Rückspiegel. Der Leib-Seele-Dualismus und das naturwissenschaftliche Weltbild*. 2006, Paderborn: mentis. 244 pages.

85. Lewis, C.I., *Mind and the World Order. Outline of a Theory of Knowledge*. Dover, New York 1991 (Nachdr.) ISBN 0-486-26564-1 ed. 1929, New York: Charles Scribner's sons. 121.

86. Lewis, L.D., P.L. Purdon, and e. al., *Rapid fragmentation of neuronal networks at the onset of propofol-induced unconsciousness*. PNAS, 2012. **109**(49): p. E3377–E3386

87. Lindemann, B., *Der philosophische Edelzwicker (II). Von Welten, Menschen und Neuronen*. Vol. 2. 2010, Homburg: invoco-verlag.

88. Lindemann, B., *Mythos wohin man sieht...* Information Philosophie, 2012. **40**(200 (3/4)): p. 172-173.

89. Lipscomb, B.W., et al., *Cell surface carbohydrates and glomerular targeting of olfactory sensory neuron axons in the mouse* J. Comp. Neurol., 2003. **467**(1): p. 22-31.

90. Lorenz, K., *Die Rückseite des Spiegels*. 1973, München: Piper. 338 Seiten.

91. Lotka, A.J., *Elements of mathematical biology (Elements of physical biology)*. 1925, New York: Dover 1956. 466.

92. Markram, H. (2013) *Human Brain Project* http://www.humanbrainproject.eu/in_brief.html **Volume**,

93. Martin, J.H., *Coding and Processing of Sensory Information*, in *Principles of Neural Science*, E.R. Kandel, J.H. Schwartz, and T.M. Jessel, Editors. 1991, Elsevier: New York. p. 329-352.

94. Melnyk, A., *A Physicalist Manifesto: Thoroughly Modern Materialism*. 2003, Cambridge: Cambridge University Press. 328 pages.

95. Metzinger, T., *Der Ego Tunnel*. 2009, Berlin: Berlin Verlag. 378 Seiten.

96. Miller, G.A., *The magical number seven plus or minus two: Some limits on our capacity to process information*. Psychol. Rev., 1956. **63**(2): p. 81-97.

97. Moore, G.E., *The refutation of idealism*. Mind, 1903. **12**: p. 433-453.

98. Mueller, P. and D.O. Rudin, *Induced excitability in reconstituted cell membrane structure*. J. Theor. Biol., 1963. **4**: p. 268-280.

99. Nagel, E., *The Structure of Science*. 1961, New York: Harcourt, Brace and World.

100. Nagel, T., *What is it like to be a bat?* The Philosophical Review, 1974. **83**(4): p. 435-450.

101. Nagel, T., *The view from nowhere*. 1986, New York: Oxford University Press. 244 pages.

102. Newman, J., B.J. Baars, and S.-B. Cho, *A neural global workspace model for conscious attention*, in *Essential sources in the scientific study of consciousness*, B.J. Baars, W.P. Banks, and J.B. Newman, Editors. 2003, MIT-Press: Cambridge, Ma. p. 1131-1148.

103. Nida-Rümelin, J., *Erregungsmuster und gute Gründe. 1. Handlungsfreiheit*, in *Zukunft Gehirn*, T. Bonhoeffer and P. Gruss, Editors. 2011, C.H.Beck: München. p. 253-257.

104. Nida-Rümelin, J. and W. Singer, *Erregungsmuster und gute Gründe: Über Bewusstsein und freien Willen*, in *Zukunft Gehirn*, T. Bonhoeffer and P. Gruss, Editors. 2011, C.H.Beck: München. p. 253-277.

105. Northoff, G., *Die Fahndung nach dem Ich*. 2009, München: Irisiana-Random House. 286 Seiten.

106. Ohm, G.S., *Die galvanische Kette, mathematisch bearbeitet*. 1827, Berlin: T.H. Riemann. 248 pages.

107. Oppenheim, P. and H. Putnam, *Unity of science as a working hypothesis*, in *Concepts, theories, and the mind-body problem*, H. Feigl, M. Scriven, and G. Maxwell, Editors. 1958, University of Minnesota Press: Minneapolis. p. 3-36.

108. Paulsson, J., O.G. Berg, and M. Ehrenberg, *Stochastic focusing: Fluctuation-enhanced sensitivity of intracellular regulation.* PNAS, 2000. **97**: p. 7148-7153.

109. Perkel and Bullock, *Neural Coding.* Neurosci. Res. Progr. Sum., 1968. **3**: p. 405-527.

110. Piaget, J., *Einführung in die genetische Erkenntnistheorie*. 1973, Frankfurt: Suhrkamp. 104 Seiten.

111. Porzig, W., *Das Wunder der Sprache*. 1950, München: Francke.

112. Poser, H., *René Descartes*. 2003: Reclam. 184.

113. Price, H. and R. Corry, eds. *Causation, physics, and the constitution of reality. Russel's republic revisited.* 2007, Clarendon Press: Oxford. 404 pages.

114. Rieke, F., et al., *Spikes. Exploring the Neural Code*. 1997, Cambridge, Mass.: MIT Press. 396 pages

115. Roth, G., *Aus Sicht des Gehirns*. 2003, Frankfurt: Suhrkamp. 214 Seiten.

116. Roth, G., *Wie einzigartig ist der Mensch? Die lange Evolution der Gehirne und des Geistes*. 2010, Heidelberg: Spektrum. 438 Seiten.

117. Russel, B., *On the notion of cause.* Proceedings of the Aristotelian Society, 1912. **13**: p. 1-26.

118. Saint-Cyr, J.A., *Frontal-striatal circuit functions: Context, sequence, and consequence.* Journal of the International Neuropsychological Society, 2003. **9**: p. 103-127.

119. Sakmann, B. and E. Neher, *Single-Channel Recording*. 1983, New York: Plenum Press. 503 pages.

120. Schmidt, S.J., *Der radikale Konstruktivismus: ein neues Paradigma im interdisziplinären Diskurs*, in *Der Diskurs des radikalen Konstruktivismus*, S.J. Schmidt, Editor. 1987, Suhrkamp: Frankfurt. p. 11-88.

121. Schutt, R. and C. O'Neil, *Doing Data Science*. 2013, Cambridge: O'Reilly. 376 pages.

122. Shannon, C.E., *The mathematical theory of communication.* Bell System Technical Journal, 1948. **27**(3): p. 379-656.

123. Shear, J., *Explaining consciousness: The hard problem*. 1997, Cambridge, MA: MIT Press. 422.

124. Shepherd, G.M., ed. *The Synaptic Organization of the Brain*. 4 ed. 1998, Oxford University Press: New York. 638 pages.

10. Bibliography

125. Simons, P., *Parts. A Study in Ontology*. 1987, Oxford: Clarendon Press, 2003. 394 pages.
126. Simons, P., *part / whole*, in *A Companion to Metaphysics*, J. Kim and E. Sosa, Editors. 1995, Blackwell: Oxford. p. 376-378.
127. Smuts, J.C., *Holism and Evolution*. Reprinted 1973 ed. 1926, Westport, CN: Greenwood Press.
128. Snow, C.P., *The two cultures*. 1969, Cambridge: Cambridge University Press. 108 pages.
129. Song, J. and B. Zipser, *Targeting of neuronal subsets mediated by their sequentially expressed carbohydrate markers*. Neuron, 1995. **14**(3): p. 537-547.
130. Szallasi, Z., J. Stelling, and V. Periwal, eds. *System modeling in cellular biology. From concepts to nuts and bolts*. 2010, MIT Press: Cambridge, Ma. 448 pages.
131. Tautz, J., *Phänomen Honigbiene*. 2007, München: Elsevier Spektrum. 278 Seiten.
132. Turing, A.M., *Computing Machinery and Intelligence*. Mind, 1950. **50**: p. 433-460.
133. Verveen, A.A. and L.J. DeFelice, *Membrane Noise*. Progress in Biophysics and Molecular Biology, 1974. **28**: p. 189-265.
134. Voit, E.O., *Computational analysis of biochemical systems*. 2000, Cambridge, UK: Cambridge University Press. 532 pages.
135. Vollmer, G., *Gretchenfragen an den Naturalisten*. 2013, Aschaffenburg: Alibri. 92 pages.
136. Weaver, R.F., *Molecular Biology*. 2011: Mcgraw-Hill Publ.Comp. 892 pages
137. Wegner, D.M., *The illusion of conscious will*. 2002, Cambridge, MA: MIT Press. 406 pages.
138. Wegner, D.M. and T. Wheatley, *Apparent mental causation. Sources of the experience of will*. American Psychologist, 1999. **54**(7): p. 480-492.
139. Whitehead, A.N., *Science and the Modern World (The Lowell Lectures, 1925)*. 1967, New York: Free Press, Macmillan Publishing Co, Inc. 212 pages.
140. Wimsatt, W.C., *Reductionism, levels of organisation and the mind-body problem*, in *Consciousness and the brain*, G. Globus, I. Savodnik, and G. Maxwell, Editors. 1976, Plenum Press: New York. p. 199-267.
141. Wimsatt, W.C., *The ontology of complex systems: levels, perspectives, and causal thickets*. Canadian Journal of Philosophy (Suppl.), 1994. **20**: p. 207-274.
142. Woodward, J., *Making Things Happen. A theory of causal explanation*. 2003, Oxford: Oxford University Press. 412 pages.

143. Woodward, J., *Causation with a Human Face*, in *Causation, Physics, and the Constitution of Reality*, H. Price and R. Corry, Editors. 2007, Oxford University Press: Oxford. p. 66-1005.
144. Ylikoski, P., *Constitutive counterfactuals and explanation.* www.timdemey.be/abstracts/yliloski.pdf, 2010.
145. Zeilinger, A., *Einsteins Spuk. Teleportation und weitere Mysterien der Quantenphysik.* 2005, München: C. Bertelsmann. 358 pages.
146. Zeilinger, A.,*Zufall als Notwendigkeit für eine offene Welt*, in *Zufall als Notwendigkeit*, H.C. Ehalt, Editor. 2007, Picus Verlag: Wien. p. 19-24.

11. Glossary

Abbreviations of rules and principles (see Chapter 8):

		Section
IDM	Idealisation in deterministic models.	2d
OSE	Over-sum effect.	2h
TDD	*Top-down* denied.	3c
USL	Universal system levels.	3f
PLC	Principle of limited cognition.	4j, 5c
MMC	Myth of the missing context.	5a
STN	Subjective transparency of the neuronal.	6a
ANA	Abstracta non agunt in concreto.	6c
CCP	Causal closure of the physical world.	6e

Abstracta are immaterial, conceptual, fictive. They have no physical extension or properties, their state-space is empty. They are not locatable in physical space, cannot interact physically, are without causal power (ANA). (Yet the neuronal representation of abstract content may be scanned neuronally without being interactively modified, attaining causal relevance). Attribution of causal power to an abstractum is a category-mistake, the fallacy of mistaken concreteness [139:51]. Abstracta have a dependent existence. They are due to a neuronal process of 'abstraction'.

Agent-world polarity: Living beings make use of internal mechanisms of control to counteract changes, exercising autonomy, agency. This property of agency draws a border between them and their environment.

Bieri-trilemma: Eliminate one wrong statement from the contradictory triple

>MdB = mind different from body
>McB = mind controls body
>CCB = causal closure of body-physical world

Suppose 'mind' means 'immaterial mind'. Then MdB and CCB are compatible, while McB is a known fallacy, that of mistaken concreteness.

Bottom-up: From a system level n up to the level above, n+1.

Causal chain, causal loop: In a model of a mechanism a sequence of physical interactions, optimized to perform the SB.

Causal closure of the (modelled) physical world (CCP): The expectation (in the idiom of level-1) that every physical change, which has a cause, is caused by physical interaction of objects. Every action on matter is interaction with other matter (a consequence of Newton's third law).

Causality-gaps: The randomness of neuronal processes was suggested to puncture the causal closure of the physical world. Through the gaps immaterial thoughts were proposed to influence material neurons [46:378].

Causal relevance: What makes a difference. R is causally relevant to the cause-effect step C → E if an independent modification of R turns C → E on or off, or modifies C. R may be an abstract content (not interacting, immaterial). Its neuronal representation (interacting material pattern) is scanned (or read) by C in a neuronal process without R being modified. (Scanning means interaction without modification of the scanned.)

Chance: On level-2 chance results from a cause lacking predictability with respect to time of effect, due to omission of level-1 information. (Section 2b, 2e).

11. Glossary

Concept: The smallest unit of thought content, and constructs of such units. Most concepts are communicated to us by peers.

Consciousness: In 1st-person perspective a multi-faceted experience of the Self. Based on neuronal processes comprising (a) activation of cortical facilities (b) evaluation of the actual sensory percepts and resulting scenes in terms of synchronic body states and (c) cogitation, prediction and flexible insightful responses:[71]

(a) *Proto-consciousness:* An activating neuronal process enabling the organism to be awake, attentive and ready to act. Mediated by the ascending reticular activating system (ARAS) of the brain stem (rhombencephalon).

(b) *Phenomenal consciousness* of the feeling Self, which attains agent-world polarity. The present outer-sensory scene is associated with the present body-percepts and qualia. Thereby the scene obtains a value with respect to felt well-being. The value may be reported to peers by behaviour, encouraging or warning them.

(c) *Access consciousness* (of humans) permits the solution of novel problems by insight into their causal structure as analysed by sub-agents working unconsciously. The solution is not of the lexical, stereotype if-then type, but synthetic by insight into laws and causes. Further, to re-experience the past by auto-biographic memory and to make predictions. Requires communication of cortical brain modules. Loss of consciousness (LOC) is caused by an interruption of this communication.

Constitutive: Composed of parts (components).

Content: What is in a container. Communicated content (abstract) is experienced as a flow of meaning. Privileged access to content is given only to the conscious Self: Its report alone can reveal meaning.

Culture is based on concepts shared by people. See C.P. Snow: "The two cultures" [128]. Notably, in terms of perspective dualism Snow's art-humanities culture is dominated by 1st-person experience while his science-engineering culture is a 3rd-person effort (Section 7c).

Emotions: Somatic states. Stereotype inbred body-reactions and beha-

---

71 The following division into phenomenal and access consciousness is due to Ned Block [19], the characterisation of feelings as markers of body states is due to Antonio Damasio [33] while work of unconscious sub-agents coordinated by a common theatre scene or work-space is due to Bernard Baars [7].

viours, consciously experienced as *feelings* [33]. Emotions 1. serve to bodily manage states like fear, anger, sadness, disgust, surprise, happiness (Darwin) and 2. serve as stereotype behaviour-signals to peers (facial expression).

Experience: Awareness of a sketch of content in the "theatre of consciousness", on the mental simulation-stage. The sketch is multimodal, involves several senses. Experience serves for communication among a group of neuronal sub-agents represented by the Self-agent.

Feelings: (my wording) 1. Awareness of body-percepts (including qualia) and their intrinsic evaluation 2. Awareness of outer-sense-percepts with associated body-percepts and their evaluation. To feel implies evaluation with respect to well-being.

Function: 1. Relates an input to an output (i/o-function). 2. The advantage gained from a (mostly biophysical) process.

Gap-less: Uninterrupted. "Gap-less sequence of interactions" or "gap-less causal chain" implies no interference from outside of physics. Compare the 'causality-gaps' of Falkenburg [46, 47].

Hard problem: According to Chalmers, how phenomenal consciousness (as raw experience) [19] arises, or how the mental arises from the physical [25].

Hardware is considered here to probe the concept of a gap-less causal chain of physical interactions, free of interference from abstract software (language). Importantly, digital computers do not 'crunch numbers' or 'process symbols'. Rather, the computer hardware guides causal chains of patterns of physical charge. The chains may be followed backward to a keyboard and a human hand. From here the causal chain is back-traced into the realm of neuronal interaction, again a gap-less sequence of physical events. A causal effect from abstract software to concrete hardware, though commonly assumed, is not found.

Holon: Arthur Koestler's term for a collection of interacting components at system level n-1, giving rise to an over-summative SB of a whole at level n, above [81].

Immaterial: Mental-immaterial: Without extension, abstract. Without physical properties and causal power, like a thought. Often "about". Physical-immaterial at level-0: not based on matter, like information, waves and fields.

Information: A quantity measured in bits (yes-or-no decisions), which removes the uncertainty resulting from a question. The semantic aspect of information is meaning (content). Information is transmitted as a pattern (symbol), to be associated with a meaning which is not transmitted but was previously learned or implemented.

Language, spoken or written, the principal way of human communication. It uses physical and neuronal objects to symbolize world objects and fictive objects, their properties and relations, and operates with the symbols following rules. In decoding language, the physical symbols (sounds, letters) are interpreted as (represented by) neuronal symbols, instances of a neuronal code. The meaning of these symbols, in turn, may be consciously experienced by associating the symbols with their denotate, using a written code-table or memory. Thus language is experienced as a flow of meaning. Language is not space-time defined, therefore an abstract. It cannot interact but talks "about" interaction. It has no causal power but may describe causal chains. It attains causal relevance through the neuronal roots of its content.

Law: A deterministic description of a generalized strong regularity.

Learning, understanding: A pattern is given a meaning. (1) A sequence of physical patterns like letters or syllables is first combined into words which are lexically associated with low-level meaning, i.e. memorized verbal concepts. (2) Then their combination to more complex concepts (like the meaning of sentences) must be 'understood'. Such comprehension requires reconstruction with syntactic rules or reconstruction with other rules of synthesis.

Life is suggested to be an over-sum system behaviour (OSE) based on a bundle of interdependent sub-processes. It may be reduced first to the level of sub-processes and then piecewise from these sub-processes to biophysical processes and finally to basal physical events.

Markov chain: Sequences of transitions with random choices between at least two ways to terminate a state.

Matter: What has mass and other classical physical properties (Extensive: energy, mass, charge, volume. Intensive: speed, temperature, pressure, density and others). Photons, being without mass, are not part of matter.

m-causation: A case of causation where the contribution of each transition, specified in a STD, is necessary but not sufficient, resulting in an *over-summative* effect (OSE), the system behaviour SB. This is caused *exclusively* by the assembly of components working in a cycle of states. The m-causation is characteristic for kinetic models of steady state mechanisms, as these necessarily have multiple transitions and at least one SB generated by a full cycle of states.

Meaning: The qualitative, semantic aspect of information, either recalled by association of a symbol with a previously learned content or synthesized following rules.

Mechanism: A physical device made of interacting components, optimized to alter its environment in a characteristic and quantitatively more or less predictable way. Designed by man or evolved in nature.

Mind: In the *first-person perspective* the domain of a Self-agent who has the conscious *experience* of agent-world polarity and of mental processes. These include the 'psychological predicates', thought about communicated mental content, the subjective experience of control and causal power and further the experience of a multimodal sketch, displayed on the 'inner theatre stage', which can be stored in autobiographic memory. In the *third-person perspective* mind is a symbol for a bundle of specific neuronal mechanisms underpinning the conscious 'mental processes'.

Mind-body problem: If the mind is without physical extension (immaterial, an abstractum) while body and neurons have extension, how can mind affect body and body mind? (See *Bieri-trilemma,* Section 6i, 6j.) Answer: Such interaction is neither possible nor required.

Model: Cluster of related hypotheses. "Artificial construction where all extraneous detail has been removed or abstracted" [121:28].

Multimodal: Multi-medial. With contributions from several senses.

Naturalisation: Reduction (e.g. of mental phenomena) to the terms of the natural sciences.

Natural Laws: Strong regularities in the behaviour of our models of world objects. Natural laws are human statements, generalized and deterministic, about regularly observed natural events, as reflected in our models. The statements are always true and are entirely dependent on our authorship, on our existence as rational human beings. Further, in these statements the natural events are grouped and ordered in a way dependent on us. However, the natural events may be reduced to basal events. These are (arguably) not dependent on us, they are independently real.

Non-physical entities: Abstracta like representation, experience, content, meaning, language, number, information, thought. Abstracta lack the space (and often time) designation of physical objects. They are without causal power because, being non-physical, immaterial, they cannot interact with physical objects, cannot encounter them in space and time.

OCA: Organized component activity. Transition. See Table 2.1.

Over-summative effect (OSE): The effect (SB) of a steady state mechanism, achieved by concerted interaction of interdependent parts and their micro-properties in a cycle. The result (SB) is a reducible, explainable macro-property, a *weak emergence* in the "innocent" sense of Dennett [38]. See Chapters 2 and 8.

Perspective Dualism: The term is used here to designate 'first-versus third-person perspectives',[72] where the first can arguably be explained by the third. I do not imply, as J. Habermas does, that the first-person or mental perspective is irreducible to the third-person or physical perspective.

Physicalism: The expectation that all phenomena can be explained physically. Implies the unity of science. *Hypothesis of reductive physicalism (RP):* Immaterial mental states and events are explainable by reduction to neuronal (physical) processes. For every immaterial phenomenon has a physical basis, a set of material micro-phenomena on which it exclusively depends. *Non-reductive physicalism (NRP):* Mental states, though caused by physical states, cannot be reduced to them, because they belong to a different ontological class. *Post-reductionism:* Even single-level reduction of the mental is not possible because the reduction base is incomplete and the mind more than a physical system.

Plan: A script describing with symbols events to be realized in the future.

Post-Reductionism: See *Reduction* or *Physicalism*.

Properly grounded: See *Universal levels*.

Psychological predicates: Apparent mental processes like thinking, feeling, remembering, judging, deciding, acting, behaving, control-

---

[72] Also known as epistemic dualism [57, 58]. For roots of this dualism, see René Descartes, quoted in [112:142], Franz Brentano [21:124], Ludwig Bertalanffy [16:95ff] and Thomas Nagel's 'dual aspect theory' [101:28].

ling, causing [14]. May be reduced to neuronal mechanisms (Section 6k).

Qualia (raw feels): The indescribable intrinsic quality of our subjective feeling-experience in phenomenal consciousness. How it feels subjectively, for example to have a headache or taste a wine or be a bat [e.g. 85, 100]. Qualia are experienced and arguably have a neuronal reduction base. - According to J. Kim, qualia are not task-oriented, therefore not functionally defined and not functionally reducible [74]. I argue that, as 'raw feels' qualia nevertheless classify as feelings and evaluate to, say, pleasant, neutral or dislikeable. Thus they are task-oriented in the sense of Damasio's markers of body states [33:55] which guide our behaviour based on inherent evaluation with respect to bodily well-being. Then qualia have a functional aspect which makes them functionally reducible to neuronal mechanisms even if the quality of the 'raw feel' cannot be verbalised.

Randomness: Lack of order, lack of predictability. In the neuronal system reliability of mechanisms appears to be instrumented, providing an additional dimension for coding. Low signal strength is associated with randomness and high strength with predictability [78]. In turn, population averaging across many parallel items (e.g. neurons) improves the overall reliability of the neuronal mechanism. Full deterministic reliability is not possible for neuronal mechanisms, as probabilistic fluctuations resume even in the deterministic limit.

Reduction: Here used in the sense of explanatory reductions of science. *Single-level reduction:* Explaining a system behaviour at level n with a mechanism or a set of micro-properties at level n-1, below. Besides its explanatory value it is a necessary check which, where successful, suggests absence of mistakes in reasoning. *Multi-level reduction:* Ranging over more than two system levels, in the end down to physical base, it amounts to a skipping of intermediate explanations, thereby hindering comprehension. Reduction is a 3rd-person endeavour, even though associated with the subjective experience of comprehension. Mental phenomena cannot be reduced within the 1st-person perspective (Section 4b), perspective must first be changed to 3rd-person (Red B, Table 4.1). *Post-Reductionism*, applied to the mental, maintains that reduction generally is not possible because the reduction base is incomplete and the mental more than a physical system. Above the reduction of life, mind, art, agency, dignity and consciousness was explored in some detail (Sections 4d, 6k, 4h, 4d, 6g, 7b).

SB: System behaviour of mechanisms, *net cycle rate*. An over-sum behaviour,[73] not explained by single transitions or addition of transition rates. Rather, steady-state SB requires the interdependence of serial transitions organised in a cycle. Single transitions of the cycle are necessary but not sufficient for the generation of SB. The 'design', the structure of the STD needs to be known for an understanding of SB.[74] This may generate a "more than transitions" intuition.

Scan: Interaction without modification of the scanned physical object, which may have causal relevance for the scanner.

Self: In the *first-person perspective* the Self is an agent possessing mind and body, positioned in the world in past, presence and future. The conscious Self views itself as separated from the world in an agent-world polarity. Further, it views itself unrelated to its own neuronal system. This comes about because we perceive only what our senses tell us, and we have no sensory organ to notice our own neuronal activity. In the *third-person perspective* the Self is a symbol for several interacting neuronal agents (bundled in a holon) engaged to solve problems by concerted action. The neuronal basis of the conscious Self seems to be the so-called executive system of the dorso-lateral prefrontal cortex and the anterior cingulate cortex [5:49,51].

Software is language. It is not space-time defined, therefore an abstract. It cannot interact but is "about" interaction. It cannot cause by itself but may describe causal chains. Its representation (interacting or scanned material pattern) may be causally relevant.

State: In finite-state systems one of several mutually exclusive combinations of component properties of finite life-time, as defined in the STD (state transition diagram). A *cycle of states* allows generation of SB of a mechanism in the steady state.

STD: State transition diagram. See Table 2.1.

Supervenience: Introduced by Donald Davidson [35]: Asymmetrical 'S

---

73 Bertalanffy termed it 'non-summative' and 'non-trivial' [17:67f]. Here 'over-summative' is preferred, in order to relate to the part / whole intuition of Aristotle: „The whole is something apart from the parts. It is more than their mere sum." [e.g. 3], commonly referred to as the 'over-sum principle'.

74 SB is a reducible (explainable) macro-property or -behaviour, a *weak emergence* in the "innocent" sense of Dennett [38]. While weak emergences are common, examples for *strong emergences*, which definitely cannot be reduced, are hard to find [11:229]

in virtue of R' or 'S necessitated by R' relation. S supervenes over R if S cannot change without a change of R. S is linked to R by identitiv, constitutive or causal relations [60:67].

Symbol: Sign representing a denotate that has an existence independent of the sign. (While the sign is dependent on the denotate: asymmetry of dependence.)

System: Model, description.

Thought: A mental operation with concepts according to syntactic rules. A try-out of action with reduced risk. Acting in virtual space. The mental experience of a flow of abstract content. Immaterial thought has no causal power. Also the neuronal mechanism underpinning this experience (has causal power).

Top-down: Effect from a system level n down to the level below, n-1. A whole at level n cannot change its components at the level below (TDD). If system levels differ only by scale (magnification, zoom, Section 3b), there can be no inter-level effect. One magnification cannot affect another.

Transparent: Present but unnoticed, like glass to vision. There is subjective transparency of our neuronal activity (STN) because we have no sensory organ for such activity [e.g. 87:240, 95:188,289, 97, 104:267, 105:95]. 'Through' the unnoticed neuronal system we sense features of world and body. This helps to establish the notion of a mind and Self existing independently from the neuronal-physical world.

Unity of science: One world, one science [e.g. 107], including the hope that the mental first-person perspective will be explained by neuronal phenomena, which are reducible to physical base. I hold that such multi-level reduction may be possible scientifically but subjectively it will carry little explanatory value since intermediate explanations are skipped. Multi-level reduction over-taxes our cognitive possibilities (PLC).

Universal system levels (USL): By axiom, changes on all system levels are due to the same basic physical events occurring synchronic on each level. Universal levels differ merely in the grouping of objects and events and the assignment of level-specific symbols to the groups, further in the level-specific context and idiom. As levels are linked by basal identity, causal or constitutive vertical linkage is not needed. There is one important caveat: the re-ordering must be based on nothing but known and documented lower-level phenomena and, in the end, basal events and their laws. Mistakes or wishful thinking (fictive objects taken for real) will result in failure to reduce. Given this provision of *proper grounding*, the principle of universal levels implies reduction of any phenomenon within the system to basic physical events. This includes the reduction of immaterial phenomena, 1st-person experiences, which are taken as 3rd-person explananda (Red B, Section 4b).

Whole: A symbol representing a holon, i.e. a group of interacting parts and their SB.

Working memory: A function of brain areas including the prefrontal and parietal cortex and part of the basal ganglia. Retains small amounts of sensory (and other) information in a form accessible to (mainly) the conscious Self, having only seconds duration and processing less than 10 items at once [5:259, 96]. This limitation (PLC) may be one good reason why we parse world phenomena into system levels, focussing attention onto few items (Figure 5.1).

12. Index

Listed are Chapters and Sections:

abstractum   1a, 6c, 6g, 6h, 6i, 6l, 9, 11
active ion-transport   4d
agency 4g
agent-world polarity   6a, 11
ANA   4b, 6c, 8, 11

Bieri-trilemma   6i, 11
bottom-up   3c, 6b, 6c, 6f, 8, 11

causal chain, causal loop   1b, 2e, 4d, 6c, 6d, 6e, 6f, 8, 11
causal closure (CCP)   4b, 6e, 6j, 6l, 8, 11
causality-gaps   6e, 11
causal models   2e, 11
causal relevance   1c, 6b, 6d, 6e, 6f, 8, 11
c-criteria   2e
chance   2e, 6e, 11
channel mechanism   2a, 5b
consciousness   7, 11
      proto-consciousness   7b, 11
      phenomenal consciousness   7b, 11
      access consciousness   7b, 11
constitutive   3b, 11
content   5a, 6b, 6c, 6f, 11
culture   7c, 11
cyclic STDs   2a, 2f, 11

deterministic models   1c, 2a, 2b, 2c, 2d, 8, 11
dignity 6g

emotions   4j, 11
ensemble-average   2a, 2c, 2d, 8

12. Index

experience 1b, 4b, 4h, 4i, 4j, 6a, 6b, 6c, 6f, 6h, 7a, 7b, 8, 11
explanatory value 4j, 5c, 6g, 8, 11

falsify 6l
feelings 4i, 6l, 7b, 11
function 4b, 4h, 4i, 8, 11
functional reduction 4h

gap-less causal chains 2e, 6c, 6d, 11
global work-space 7a

hard problem 7, 11
hardware 6d, 11
hierarchy of neuronal mechanisms 3a, 5b
holon 3a, 3e, 5b, 11

idealisations 2b, 2d, 11
IDM 2d, 8, 11
information 2b, 5a, 6c, 8, 11

language 6f, 7a, 11
learning 5a, 6d, 11
life 4g, 11
linkage of levels 3a, 3b, 3c, 3e, 3f, 5b, 6f, 11

macroscopic models 2b, 2d
magnification, scale, zoom 3b
Markov chain 2a, 2e, 5d, 11
m-causation 2f, 3e, 8, 11
meaning 5a, 6c, 6d, 11
mechanism 1b, 1c, 2, 11
mind 1a, 1b, 6, 7, 8, 11
mind-body problem 1a, 6i, 6j, 8, 11
mind immaterial 1c, 4b, 4f, 4h, 6c, 6e, 6f, 6h ff, 11
missing context 5a, 8, 11
mistaken concreteness 1a, 6c, 6e, 6i, 8, 11
MMC 5a, 8, 11
molecular models 1c, 2a, 2c, 2d, 4d, 5b, 11
multimodal 4i, 6h, 7a, 11

natural laws 4, 4b, 11
neuronal code 5a, 11
neuronal identity and targeting 5a, 11
neuronal mechanisms 5, 11
non-physical entities 6e, 11

OCA 2a, 11
over-sum effect (OSE) 2f, 4g, 4j, 11

pacemaker neuron 4d

perspective dualism 1a, 7, 8, 11
physicalism 1a, 1c, 4f, 6l, 8, 9, 11
PLC 4j, 5c, 8, 11
post-reductionism 1a, 4b, 4f, 4j, 5c, 8, 9, 11
pragmatic basic level 3d
probabilistic models 2a, 2c, 2d, 6e, 8, 11
proper grounding 3f, 11
psychological predicates 6g, 6h, 6k, 11
psychology, sociology etc. 6g

qualia 1a, 4i, 6l, 11

randomness 4d, 5d, 6e, 11
recall a number 6k
reducing consciousness 7b, 11
reducing dignity 6g, 11
reducing the mind 6k, 11
reduction 1, 3f, 4, 5c, 6g, 6k, 7b, 11
reduction by material synthesis 4e, 7b, 8
reduction of agency 4g
reduction of a work of art 4j
reduction of life 4g
reduction to a mechanism 4d
reduction to a model 4c
reductive modelling 4
reductive physicalism RP 1a, 1c, 6i, 6l, 8, 9, 11
relation of world and mind 6f

SB 2a, 2c, 2f, 11
scale, magnification, zoom 3b
scan 7a, 11
Self 1b, 6a, 6h, 6k, 7a, 7b, 11
software 6d, 11
STD 2a, 2f, 11
STN 4b, 6a, 8, 11
supervenience 1a, 6b, 6l, 11
switching of levels 3d
symbol 3d, 3e, 4f, 5a, 5b, 5c, 6b, 6d, 6g, 11
systems of symbols 3e

thought 2, 64, 67f, 131
thoughts control neurons? 1a, 6c, 6e, 11
TDD 3c, 11
time-average 2b, 2d, 4a
time requirement 2c, 6f, 7a
top-down 3c, 6c, 11
transparent 6a, 6c, 6f, 7a, 8, 11

undifferentiated encoding 5a
unity of science 1a, 3a, 4j, 11

12. Index

universal system levels   1c, 3f, 5c, 6f, 8, 11
USL   3f

whole   2f, 3a, 3b, 3c, 3d, 3e, 11
word of caution   5b
working memory   5b, 5c, 11

zoom   3b